PYROSEQUENCING: INNOVATIONS AND APPLICATIONS

PYROSEQUENCING: INNOVATIONS AND APPLICATIONS

Edited by:

Dr Prerna Pandey

Delve
PUBLISHING

www.delvepublishing.com

Pyrosequencing: Innovations and Applications

Prerna Pandey

Delve Publishing
2010 Winston Park Drive,
2nd Floor
Oakville, ON L6H 5R7
Canada
www.delvepublishing.com
Tel: 001-289-291-7705
001-905-616-2116
Fax: 001-289-291-7601
Email: orders@arclereducation.com

ISBN: 978-1-77361-150-1 (Hardcover)

Library of Congress Cataloging-in-Publication Data

A CIP record for this book has been applied for and is available from Library and Archives Canada.

Delve Publishing publishes wide variety of books and eBooks. For more information about Delve Publishing and its products, visit our website at www.delvepublishing.com.

DECLARATION

ABOUT THE EDITOR

Dr Prerna Pandey, a biotechnologist with several years of wet lab research experience, worked at International Center for Genetic Engineering and Biotechnology, New Delhi. She has also worked at Transasia Biomedicals and Advance Enzyme Technologies as a scientist. Prerna has published papers in peer-reviewed journals, and has submitted a number of annotated Geminiviral genome sequences in the GenBank, including two novel ones. Now, she works as a freelance scientific editor and writer.

TABLE OF CONTENTS

LIST OF CONTRIBUTORS

Prerna Pandey

Dr Prerna Pandey, a biotechnologist with several years of wet lab research experience, worked at International Center for Genetic Engineering and Biotechnology, New Delhi. Her field of research in PhD included isolation and molecular characterization of geminiviruses, genome sequencing, gene annotation, and gene silencing using the RNA interference technology. She has also worked at Transasia Biomedicals and Advance Enzyme Technologies as a scientist. Prerna has published papers in peer-reviewed journals, and has submitted a number of annotated Geminiviral genome sequences in the GenBank, including two novel ones. She has also completed her editing and proof-reading courses from Society for the promotion of Editors and Proof-readers, UK and now works as a freelance scientific editor and writer. When Prerna is not busy with her assignments, she loves to explore historical places.

Deepti Mittal

Deepti Mittal, PhD is a Scientist, with 10 years of research experience in Advanced Molecular Biology. She worked as a senior research fellow at International Center for Genetic Engineering and Biotechnology, New Delhi and earned her doctorate in Biotechnology from Banasthali University, Rajasthan. During PhD, she identified novel miRs and studied miRNA-target interaction. She specializes in next generation sequencing driven assembly, profiling and characterization of genome. Currently, she is working in Synthetic Genomics, La Jolla, CA, USA and a part of manufacturing operations team (Gene Synthesis and Cell Engineering), manufacture genes and genome segments, using proprietary methods and a creative Molecular Biology approach. Previously she also worked at ThermoFisher Scientific, Carlsbad, CA, USA, as Scientist in NGS development team.

Neha Sharma

Dr. Neha Sharma has obtained her Ph.D. from International Centre for Genetic Engineering and Biotechnonolgy, New Delhi. She has pursued her Ph.D. work in functional characterization of rice miRNAs in development and stress. She also has patent for a agronomically important phenotype for a rice transgenic. She has several peer-reviewed articles in journals of international repute and received several awards in National and international conferences. Currently, she is working as an assistant professor in department of biotechnology, Guru Ghasidas University, a central university, Bilaspur, Chattisgarh, India.

Shiv Sanjeevi Sripathi

Shiv Sanjeevi Sripathi completed his BSc in Biotechnology from Mumbai University and secured second rank in Mumbai University and first rank in his college: Kishinchand chellaram College. Shiv secured first rank in his college KET's V.G.Vaze College and has worked on a stem cell project at the Specialized Centre for Cell Based Therapy (SCCT), KEM Hospital at Mumbai, following which he was a Junior Research Fellow at Microbiology & Cell Biology Department, Indian Institute of Science, Bangalore from August 2009 to November 2011. The field of work involved cloning of cell wall genes and transcription factors in E. coli & M. smegmatis. This was followed by a stint as a Research and Development Executive at Transasia Biomedicals, Mumbai where he worked on patent writing and Immunoassays from 2011 till 2013. In 2013, he entered the teaching field where he has been tutoring biology and chemistry to the students of 11th and 12th grades along with English grammar and communication skills to people of all age groups. Shiv also loves helping people write scientific manuscripts.

Santiago Català

Instituto Agroforestal Mediterráneo, Universitat Politècnica de València, Camino de Vera s/n, Valencia, Spain,

Ana Pérez-Sierra

Forest Research, Alice Holt Lodge, Farnham, Surrey, United Kingdom
Instituto Agroforestal Mediterráneo, Universitat Politècnica de València, Camino de Vera s/n, Valencia, Spain,

Paloma Abad-Campos

Instituto Agroforestal Mediterráneo, Universitat Politècnica de València, Camino de Vera s/n, Valencia, Spain,

Elham A. Kazeeroni

Department of Crop Sciences, College of Agricultural and Marine Sciences, Sultan Qaboos University, Seeb, Oman

Abdullah M. Al-Sadi
Department of Crop Sciences, College of Agricultural and Marine Sciences, Sultan Qaboos University, Seeb, Oman

G. Fusco
Department of Animal Health, Experimental Zooprophylactic Institute of Southern Italy, Via Salute, 2, 80055 Portici (NA), Italy

M.G. Amoroso
Department of Animal Health, Experimental Zooprophylactic Institute of Southern Italy, Via Salute, 2, 80055 Portici (NA), Italy

N. Gesualdi Montesano
Qiagen Srl, Via Grosio 10/8, 20151 Milano, Italy

M. Viscardi
Department of Animal Health, Experimental Zooprophylactic Institute of Southern Italy, Via Salute, 2, 80055 Portici (NA), Italy

Fabrice Armougom
URMITE - UMR CNRS 6236, IRD 3R198, Université de la Méditerranée, Faculté de Médecine, 27 Boulevard Jean Moulin, 13005 Marseille, France

Didier Raoult
URMITE - UMR CNRS 6236, IRD 3R198, Université de la Méditerranée, Faculté de Médecine, 27 Boulevard Jean Moulin, 13005 Marseille, France

Shavanthi Rajatileka
Centre for Research in Biosciences, Department of Biological, Biomedical and Analytical Sciences, Faculty of Health and Applied Sciences, University of the West of England, Bristol BS16 1QY, UK.

Karen Luyt
Neonatal Neuroscience, School of Clinical Sciences, University of Bristol, St Michael's Hospital, Southwell Street, Bristol BS2 8EG, UK.
Regional Neonatal Intensive Care Unit, St Michael's Hospital, University Hospitals NHS Trust, Bristol BS2 8EG, UK.

Maggie Williams
Bristol Genetics Laboratory, Pathology Sciences, Blood Sciences and Bristol

Genetics, Southmead Hospital, Bristol BS10 5NB, UK.

David Harding
Regional Neonatal Intensive Care Unit, St Michael's Hospital, University Hospitals NHS Trust, Bristol BS2 8EG, UK.

David Odd
Neonatal Neuroscience, School of Clinical Sciences, University of Bristol, St Michael's Hospital, Southwell Street, Bristol BS2 8EG, UK.
Neonatal Intensive Care Unit, Southmead Hospital, North Bristol NHS Trust, Bristol BS10 5NB, UK.

Elek Molnár
School of Physiology and Pharmacology University of Bristol, Medical Sciences Building, University Walk, Bristol BS8 1TD, UK

Anikó Váradi
Centre for Research in Biosciences, Department of Biological, Biomedical and Analytical Sciences, Faculty of Health and Applied Sciences, University of the West of England, Bristol BS16 1QY, UK.

Zongli Zheng
Department of Medical Epidemiology and Biostatistics, Karolinska Institutet,

Abdolreza Advani
Swedish Institute for Infectious Disease Control, Solna,

O¨ jar Melefors
Department of Microbiology, Tumor and Cell Biology, Karolinska Institutet, Stockholm and
Swedish Institute for Infectious Disease Control, Solna,

Steve Glavas
Swedish Institute for Infectious Disease Control, Solna,

Henrik Nordstro¨ m
Department of Microbiology, Tumor and Cell Biology, Karolinska Institutet, Stockholm and
Swedish Institute for Infectious Disease Control, Solna,

Weimin Ye
Department of Medical Epidemiology and Biostatistics, Karolinska Institutet,

Lars Engstrand
Swedish Institute for Infectious Disease Control, Solna,
Department of Microbiology, Tumor and Cell Biology, Karolinska Institutet, Stockholm and

Anders F. Andersson
Swedish Institute for Infectious Disease Control, Solna,
Limnology/Department of Ecology and Evolution, Evolutionary Biology Centre, Uppsala University, Uppsala, Sweden

Ana V. Almeida
KIST-Europe, Microfluidics Group, Campus E7.1, 66111 Saarbrücken, Germany.
Mechatronics Department, Universität des Saarlandes, Campus A5, Saarbrücken, Germany
Central European Institute of Technology (CEITEC), Brno University of Technology (BUT), Technická 3058/10, Czech Republic

Andreas Manz
KIST-Europe, Microfluidics Group, Campus E7.1, 66111 Saarbrücken, Germany.
Mechatronics Department, Universität des Saarlandes, Campus A5, Saarbrücken, Germany
Central European Institute of Technology (CEITEC), Brno University of Technology (BUT), Technická 3058/10, Czech Republic

Pavel Neužil
KIST-Europe, Microfluidics Group, Campus E7.1, 66111 Saarbrücken, Germany.
Department of Microsystem Engineering, School of Mechanical Engineering, Northwestern Polytechnical University (NPU), 127 West Youyi Road, Xi'an Shaanxi, PR China
Central European Institute of Technology (CEITEC), Brno University of Technology (BUT), Technická 3058/10, Czech Republic

Lin Liu
NGS Sequencing Department, Beijing Genomics Institute (BGI), 4th Floor, Building 11, Beishan Industrial Zone, Yantian District, Guangdong, China

Yinhu Li
NGS Sequencing Department, Beijing Genomics Institute (BGI), 4th Floor, Building 11, Beishan Industrial Zone, Yantian District, Guangdong, China

Siliang Li, Ni Hu
NGS Sequencing Department, Beijing Genomics Institute (BGI), 4th Floor, Building 11, Beishan Industrial Zone, Yantian District, Guangdong, China

Yimin He
NGS Sequencing Department, Beijing Genomics Institute (BGI), 4th Floor, Building 11, Beishan Industrial Zone, Yantian District, Guangdong, China

Ray Pong
NGS Sequencing Department, Beijing Genomics Institute (BGI), 4th Floor, Building 11, Beishan Industrial Zone, Yantian District, Guangdong, China

Danni Lin
NGS Sequencing Department, Beijing Genomics Institute (BGI), 4th Floor, Building 11, Beishan Industrial Zone, Yantian District, Guangdong, China

Lihua Lu
NGS Sequencing Department, Beijing Genomics Institute (BGI), 4th Floor, Building 11, Beishan Industrial Zone, Yantian District, Guangdong, China

Maggie Law
NGS Sequencing Department, Beijing Genomics Institute (BGI), 4th Floor, Building 11, Beishan Industrial Zone, Yantian District, Guangdong, China

PREFACE

Pyrosequencing is a DNA sequencing technique that involves sequencing by synthesis, and employs a series of enzymes to detect nucleic acid sequences during the synthesis and enables qualitative analysis of the DNA. A major advantage of this technique is that the labeled primers and nucleotides, as well as gel-electrophoresis is not required. The method requires minimal amounts of samples and gives a high throughput at a lower cost. Besides, the sequence reaction time per base calling is shortened, with higher accuracy, better automation and parallel processing. Pyrosequencing offers a valuable tool for obtaining and analysis of accurate data, identification of new alleles, polymorphic genetic markers, especially Single Nucleotide polymorphisms.

The primary purpose of this book is to discuss basics of the Pyrosequencing technology and its key features, including the general sequencing modes, limitations, and future applications. We look forward to feedback by our readers

Editor

CHAPTER 1

FASCINATING WORLD OF NEXT GENERATION SEQUENCING: AN OVERVIEW AND ITS IMPACT ON GENOMICS

Deepti Mittal [1], Neha Sharma[2]

[1]Deepti Mittal, PhD is a Scientist, with 10 years of research experience in Advanced Molecular Biology. She worked as a senior research fellow at International Center for Genetic Engineering and Biotechnology, New Delhi and earned her doctorate in Biotechnology from Banasthali University, Rajasthan. During PhD, she identified novel miRs and studied miRNA-target interaction. She specializes in next generation sequencing driven assembly, profiling and characterization of genome. Currently, she is working in Synthetic Genomics, La Jolla, CA, USA and a part of manufacturing operations team (Gene Synthesis and Cell Engineering), manufacture genes and genome segments, using proprietary methods and a creative Molecular Biology approach. Previously she also worked at ThermoFisher Scientific, Carlsbad, CA, USA, as Scientist in NGS development team.

[2]Dr. Neha Sharma has obtained her Ph.D. from International Centre for Genetic Engineering and Biotechnonolgy, New Delhi. She has pursued her Ph.D. work in functional characterization of rice miRNAs in development and stress. She also has patent for a agronomically important phenotype for a rice transgenic. She has several peer-reviewed articles in journals of international repute and received several awards in National and international conferences. Currently, she is working as an assistant professor in department of biotechnology, Guru Ghasidas University, a central university, Bilaspur, Chattisgarh, India.

INTRODUCTION

Although in the past decade, the genome of several species was sequenced and studied by using several different platforms of sequencing technology, but in the era of personalized medicine and complex genome, imploration for development of new technologies is rational. This rapid change in sequencing technology is giving a path to scientist for analysing and understanding the genome. This advancement has also opened the way in several fields like comparative genomics, Biomedical research, personal genome projects. Due to this advancement, it is possible to analyse complete genome [1] and large gene panels[2]. In biomedical research, it is being helpful in the detection of various genomic alterations such as single nucleotide mutations, small insertions and deletions, chromosomal rearrangements, copy number variations[3]. There is huge change in the field of personalized medicine that impact human health by providing accurate diagnosis, prognosis and consultancy for treatment [3,4].

Particularly generations of sequencing technologies are distinguished in several categories as First Generation Sequencing, Second Generation Sequencing and Third Generation Sequencing (still in development).

The overall aim of this chapter is to provide an overview of the advances of these technologies, the bioinformatics tool, and the application of NGS in today's world.

SEQUENCING PLATFORMS

The potential to determine the sequence of DNA or RNA is essential to biological and medical research because this information contain hereditary and biochemical properties of global life. Over the last five decades there is drastic change in this technology starting from sequencing short oligonucleotides to millions of bases, from coping towards the deduction of the coding sequence of one gene to whole genome sequencing. There are three different generations of sequencing technology to address the problem of how to sequence DNA.

Timeline of Sequencing Technology:

1972: First gene from RNA was sequenced by Walter Fiers.

1976: First complete genome was sequenced by Fiers.

1977: A new method for sequencing DNA was discovered by Maxam Gilbert.

1977: DNA sequencing with chain termination inhibitors was discovered by Sanger.

1987: Applied Biosystems released first time ABI 370 (first automated sequencing machine).

1990: large-scale sequencing trials were started by National Institutes of Health (NIH).

1995: Craig Venter at TIGR published *Haemophilus influenzae* genome.

2000: The full genome sequence of the model organism Drosophila melanogaster (fruit fly) was completed.

2000: Brenner and Lynx Therapeutics published "MPSS", massive parallel bead-base sequencing.

2002: The genome of parasite *Plasmodium falciparum* was published.

2003: HGP (Human Genome Project) "complete" genome was released.

2004: Pyrosequencer was released.

2006: Genome Analyzer was launched by Illumina.

2007: Applied Biosystems released Sequencing by Oligonucleotide Ligation Detection (SOLiD).

2009: Pacific Biosciences introduced Single Molecule Real Time (SMRT) DNA Sequencing technique.

2010: ThermoFisher Scientific launched the first non-optical system.

2011: From ThermoFisher and Illumina, 2 sequencing platforms were launched named as Personal Genome Machine (PGM), MiSeq respectively.

2013: The Zebrafish genome sequence was completed.

2014: Illumina launched its HiSeq X Ten system, consisting of 10 ultra-high-throughput sequencers.

First Generation Sequencing

In 1953, Watson and Crick solved three-dimensional structure of DNA, by working on crystallographic data produced by Franklin and Wilkins [5,6]. Initially sequencing effort was focused on most easily available populations of pure RNA species, such as microbial rRNA or tRNA, or the genomes of ssRNA bacteriophages.

Sanger and coworkers also developed another sequencing technique which was based on the detection of radio labelled partial-digestion fragments after two-dimensional fractionation[7]. Due to this development, researchers could grow pool of ribosomal and transfer RNA sequences [8,9].

Then after a time being, two-dimensional fractionation was replaced by polyacrylamide gel due to its capability of greater resolving power. By using this polyacrylamide gel, researchers could determine the nucleotide position. Sanger and colleagues sequenced first DNA genome, bacteriophage φX174 by using this technique[10].

Maxam and Gilbert used this polyacrylamide gel approach but that was different from Sanger. Radiolabeled DNA was treated with chemicals which could break chain at specific bases, after running on polyacrylamide gel the length of cleaved fragments can be determined and therefore sequence inferred. This technique was widely adopted first time and considered as first-generation DNA sequencing.

Another breakthrough of DNA sequencing technology was developed by Sanger in 1977 by his chain-termination or dideoxy technique[11]. In this method, chemical analogues of dNTPs are used. The addition of chain-terminating dideoxynucleotides was very selective in this process and that was held by DNA polymerase during in vitro DNA replication. This technology was one of the most common sequencing technology due to its accuracy, robustness and ease. Several changes were made to improve this technology further as use of fluorometric based detection and improved detection by capillary based electrophoresis. Both improvements were very helpful in the development of automated DNA sequencing machines[12-15].

Then dideoxy sequencers developed by using polymerase chain reaction (PCR)[16,17] and recombinant DNA technologies [18], which allowed simultaneous sequencing of hundreds of samples. This sequencer was used for Human Genome Project [19].

Second Generation Sequencing: Next Generation Sequencing Technologies (NGS)

In last decade, Next-generation sequencing (NGS), was developed. This technology allows simultaneous sequencing of millions of DNA fragments. In this revolutionary era, this advanced technology really helped researched in tremendous way. Traditional sequencing had its own limitations as due to the high costs and much wok, it was only performed on specific DNA regions and for specific samples. With NGS, a complete genome can be sequenced in a few days for less than $1000 per genome. We are seeing the progress and impact of NGS in understanding biological processes of diseases and in personalized medicine.

The basic difference between traditional Sanger sequencing and NGS is in several ways. A very first thing is, DNA sequencing libraries are clonally amplified *in vitro*. Second, DNA is sequenced by synthesis. And then, amplified DNA templates are sequenced simultaneously in a massively parallel fashion and this step doesn't require a physical separation step. Though there are several high-throughput sequencing platforms available but they all are slightly different either in terms of chemistry or strategy.

Based on application, today NGS technology is grouped in two main types that depends on short reads and long reads. For short read sequencing (shorter than 300 bp), researchers usually use Illumina platform that is based on "sequencing by synthesis" (SBS) [20]. Another reliable and short read sequencer is based on semiconductor technology from Ion Torrent [21]. For long read sequencing, mainly PacBio or Roche is in use which is known as "single molecule real-time" (SMRT) technology and useful for longer than 2.5 Kb [22, 23]. There is one more technology in market, Nanopore Technology which uses single stranded pore. It allows to sequence very long molecules (>10 Kb) with relative low cost but its shortcoming is also there ie higher error rate compared with other sequencers[24]. SMART and Nanopore technologies has been categorized as Third Generation Sequencing. Earlier 454 Pyrosequencing was in use for long read sequencing. This technology is in use now a days but as time is rapidly changing, people are moving towards new technologies with more precision and accuracy.

NGS platforms which are commonly available and in use:

- 454 Pyrosequencing
- Sequencing by ligation (SOLid)
- Illumina
- Ion Torrent

454 Pyrosequencing

454 Genome Sequencer was developed in 2005 by Rothberg and colleagues [25]. This is a sequence-based detection technology that enables rapid and accurate quantification of sequence variation. This technology relies on the detection of pyrophosphate release when nucleotides are incorporated during a sequencing reaction. There are four enzymes which are involved in this process. A luciferase is an enzyme which emits light in the presence of ATP. During replication, each addition of a dNTP releases pyrophosphate then PPi allows adenosine phosphosulfate (APS) to be converted to ATP; this ATP allows the conversion of luciferin in to oxyluciferin which generates visible light. This light is detected by sensor and het converted into a peak as a raw data. If dATP is added in the reaction and light is emitted, your sequence starts with A. If not, the dATP is degraded by Apyrase enzyme. In this method, the height of each peak is proportional to the number of nucleotides incorporated. Sequential addition of nucleotides allows quantitative decoding of the sequence to analyze [26].

Pyrosequencing technology has been successfully used in several areas. It has been used to correlate DNA methylation with tumor type and gene expression, to measure cellular response to treatment with demethylating agents. Also, it has been very helpful to detect the changes in methylation state in relation to genetic imprinting, tumorigenesis and exposure to environmental toxins. This sequencing platform has been proved precise and accurate in sensitive mutational analysis or in quantifying alleles from mixed cell populations. Another important use of this technology is in genetic testing. Pyrosequencing delivers sequence information, so it is so helpful in the evaluation of genetic variation such as insertion, deletions, SNPs, short tandem repeats, and variable gene copy number. It is possible to assay several contiguous sequence variants in a single run.

For microbial and viral pathogens, it's always important to diagnose accurate and specific typing [27]. For bacterial identification, this technology was developed as a method which was suitable for regions other than 16s rRNA of different bacteria [27]. The use of pyrosequencing was evaluated by Bao et al., for the analysis of acid-fast bacilli (AFB) isolates. This study showed the comparative study with phenotypic analysis and Sanger sequencing. Pyrosequencing analyzed 114 (98%) of 117 AFB isolates tested and correctly identified [28]. Other report showed that a panel of 32 genes was identified which was associated with prognosis of cervical squamous cell carcinoma. *VIM* gene methylation in cervical squamous cell carcinoma was shown to predict a favorable prognosis. Also, a three-marker panel (*SOX1*, *IRAK3*, and *L1-MET*) was validated using pyrosequencing that can predict between bladder cancer recurrence with sensitivity of 93% and specificity of 94% [29].

In applied research, it has been used to detect lyssavirus by targeting on the 3′ terminus of the N gene. This technology has been used in clinic for prognostic purposes. It has been used for newly diagnosed glioblastoma, although some uncertainties remain regarding methodologies, cut-off definitions, and optimal use in the clinical setting. Though for long read length, it is most effective platform but still it has its own limitation. The limitation is that it frequently misidentifies the length of homopolymers. Also, it is cost ineffective compared with several other NGS platforms. Even in each limited budget, researcher can produce more sequences with Illumina, SOLiD, or Ion Torrent.

Sequencing by ligation (SOLid)

A revolutionary new platform that delivers ultra-high throughput with unmatched accuracy. The robustness and flexibility of SOLid system enable to wide variety of applications including WGS, targeted re-sequencing, gene expression, small RNA analysis and chromatin immunoprecipitation. The workflow of each application differs only in the upfront sample preparation and downstream data analysis step. The sequencing chemistry and imaging steps remain constant. The SOLid system allows researchers two choices of sample prep method, fragment library or mate-paired library. In both cases target DNA is sheared to a specific size and adapters are ligated to the ends of the resulting fragments. A fragment library incorporates a single piece of DNA whereas mate pair library incorporates 2 DNA sequences they are known distant apart in the target samples. The resulting libraries contain millions of unique molecules representing the entire target sequence. Each molecule is then clonally amplified on to beads in an emulsion PCR reaction. The sample is enriched for beads containing library template yielding greater than 90% positive bead count. Beads are covalently attached to the glass slide that allows the flexibility to analyze 1, 4 or 8 samples on single slide. With barcoding one can pool up to 16 libraries, further increasing the throughput to 256 samples in a single run. On the SOLid analyzer, the templated beads are combined with universal sequencing primer, ligase and large pool of di-base probes. The di-base probes are fluorescently labelled with four dyes and each dye represent 4 of 16 possible dye nucleotide sequences. The complementary probe hybridizes to the template sequence and is ligated. After fluorescence is measured the dye is cleaved off leaving a free 5' phosphate group available for further reactions. This process is repeated with 7 cycles and yield of 35 bp reads. Additional cycles can be added to extend the read length. The synthesized strand is removed, a new primer is hybridized offset by one base and ligation cycles are repeated. This primer re-set process is repeated for 5 rounds providing a dual measurement of each base separated by several cycles which increases the sequencing accuracy. With two independently controlled flow cells, this system allows researchers to run two completely different experiments in a single or staggered runs increasing the productivity. Furthermore, increasing read length will deliver additional throughput advances. This system delivers the robustness and reliability with features like real time run stages and the ability to reenter the workflow at multiple points, downtime is minimized [30]. Though this technology is in use to sequence nucleic acid with its great throughput but it has great advantage in transcriptomics and epigenetics.

By using these high throughput sequencing platform, it is possible to study the complexity of cancer. NGS methods has made it easy to study detection of genomic alterations, substitutions, insertions, deletions and copy number alterations. By using this SOLid sequencing, WGS was performed on FFPE samples of prostate cancer [31]. There is one more report in which by exome sequencing of xenograft tissue, mutation frequency was studied in cancer [32]. Though this technology was used in diverse areas but due to high cost and data analysis challenges the use was limited in the field. This had limitation in using FFPE samples as it was difficult to maintain the high-grade DNA integrity of this tissues. The major disadvantage of this technology is the very short sequencing reads generated.

To overcome these issues and to study FFPE samples, Illumina platform has been used in wide varieties and still expanding with new applications every other day [33,34].

Illumina Sequencing

Illumina sequencing technology allows a wide variety of applications, that allow researchers to ask almost any question related to genome, transcriptome, or epigenome of any organism. Next-generation sequencing (NGS) methods are different based on their principle that include library preparation and data analysis. Illumina launched Genome Analyzer first time in 2006 and it became famous among all researchers due to its larger amount of data in a more cost-effective manner. Over almost a decade, read lengths have increased and Illumina platform has become most popular [20].

Sequencing by synthesis utilizes the step by step incorporation of reversibly fluorescent and terminated nucleotides. This system is able to overcome the disadvantages of pyrosequencing system by only incorporating single nucleotide at a time. However, as a sequencing reaction proceeds the error rate of the machine also increases. This is due to incomplete removal of fluorescent signal which leads the higher background of noise level.

Illumina workflow of sequencing has 4 basic steps including library preparation, cluster generation, sequencing and data analysis [30]. This platform performs massively parallel sequencing of hundreds of millions of fragments of DNA. First the DNA is fragmented, sheared ends are repaired and adenylated. Adapter oligoes are ligated to both ends of the fragments. These fragments are size selected and then purified. Cluster generation, single molecules are isothermally amplified in a flow cell to prepare them for high throughput sequencing. The flow cell has a dense number of oligoes grafted to its surface, these oligoes bind to the adapters ligated to the library fragments. Single DNA molecules hybridize to the lawn of oligoes and bound fragments are extended to create copies. These copies are covalently bound to the flow cell surface. Each library fragment is clonally amplified through a series of extensions and isothermal bridge amplifications resulting in hundreds of millions of unique clusters. The reverse strands are cleaved and washed away. Ends are blocked and the sequencing primer is hybridized to the DNA templates. After cluster generation, library is ready for sequencing. In sequencing, hundreds of millions of clusters are sequenced simultaneously. DNA templates are sequenced base by base in parallel using 4 fluorescently labelled reversibly terminated nucleotides. All 4 bases compete to bind to the template. This competition ensures the highest accuracy. After each round of synthesis, the clusters are excited by a laser and emitting a color that identifies a newly added base. The fluorescent label and blocking group are then removed and allowing for the addition of the next base. This method reads the single base in each cycle enable accurate sequencing through difficult regions such as homopolymers and repetitive sequences. As described earlier, the first Illumina sequencer was launched in 2006. This Genome analyzer was improved in terms of performance in subsequent series GAIIX series, released in 2009. Then in 2010, HiSeq 2000 was launched, chemistry was same as GA with astonishing improvements in terms of data output and accuracy. In 2011, MiSeq was released which shared most technologies with HiSeq. MiSeq is very easy to use, in use and especially convenient for amplicon and bacterial sample sequencing [35].

Although Illumina NGS platform is still growing with new and advanced applications, but it has been very important and fruitful technology so far. In the field of genome, transcriptome or epigenome of any organism, Illumina NGS method easily allow researchers to ask almost each question.

Ion Torrent

Ion torrent technology takes an entirely new approach to NGS, making it massively scalable, faster, simpler and more affordable than ever before. Sequencing of DNA is done using a semiconductor chip like the chip found in digital camera. While the chip in camera has a sensing layer covered with millions of pixels that translates light in to digital information and Ion chip has millions of wells covering that pixels. These wells capture chemical information from DNA sequencing and translates in to digital information or base call.

Sequencing process starts when a sample of DNA is cut up in to millions of fragments each frag then attaches to its own beads and its copied until it covers the beads. This automated process covers millions of beads with millions of different fragments. These beads then flow across the chip, each depositing in to a well. Then the chip is flooded with one of the four DNA nucleotides. Whenever a nucleotide is incorporated into single strand of DNA, H+ ion is released. This is how the Ion Torrent system sequences DNA by reading this chemical change directly on the chip. The H+ changes the pH of the solution in the well and ion sensitive layer beneath the well measures that change in pH and convert it in to the voltage. This voltage change is recorded and indicates that nucleotide is incorporated and the base was called. The process is repeated every 15 seconds with different nucleotide washing over the chip. For example: Cytosine, a polymerase incorporates the C nucleotide in the DNA strand if the complementary G molecule is present. If the nucleotide is not complementary to the next base, no ion is released, no change in voltage and no base is called. If there are 2 identical bases next to each other, 2 nucleotides are incorporated then voltage gets doubled and chip records 2 bases called. This process happens simultaneously in millions of wells so regardless weather one million or one billion wells are used in chip, it takes only few hours [30].

Ion torrent has three different platforms, S5, Personal Genome Machine (PGM) and Proton based on its semiconductor technology. The basic difference between these platforms is the size of sample-deposition surface and distance between wells. They have different capacities in terms of maximum output of raw data that depends that which size of chip is being used. Ion torrent platform uses clonal amplification for library preparation that included emulsion PCR using non-paramagnetic beads. The template gets enriched and then gets deposited on a micro well surface. Each micro well lodges only one bead and place above the sensor plate which is sensitive to change in pH.

As Ion Torrent is giving alternative method to achieve vast range of read length, this is one of the best suited technology for small scale applications like amplicon sequencing, microbial genome sequencing, transcriptome sequencing. [36]. Recently, in June 2017, ThermoFisher has developed Oncomine Dx Target Test by using PGM platform. The assay is based on Ion AmpliSeq technology, which can screen tumor samples for multiple genetic markers using 10 nanograms of nucleic acid. It has been reported recently in Human Genomics that PGM system has been used to analyze *BRCA1* and *BRCA2* genes and it has detected point mutations and indels by using NGS [37].

Application of NGS

Through the use of high throughput NGS technologies large scale molecular data is generated which can be put forth to several applications based on the demand. The huge 'informatics' data generated from living organisms find diverse application areas such as de novo gene sequencing, targeted discovery of structural gene rearrangements and polymorphisms, disease predictive models and clinical diagnostics, gene expression analysis and so on [38, 39, 40]. Once the sequencing is done, it can be put to analysis to answer several related queries to find meaningful predictions to confirm through experimental analysis. Human genome research has gained the most from NGS as resequencing is possible in a quick, reliable and comparatively cheap manner. for example, Roche 454 technology can provide 7.5x genome coverage whereas it is 20x using Illumina genome analyzer [41,42]. The whole genome genotyping approach on HT-NGS effectively enables unlimited multiplexing and unconstrained single nucleotide polymorphism (SNP) selection, for example typing of HLA genotypes in human [43] and genome-wide fetal genotyping using non-invasive HT-NGS of the mother's blood [44]. NGS had played a pivotal role in providing genotype-phenotype associations for living organisms ranging from microbes to plants, animals and humans [45]. NGS had also led to the discovery of novel occurrence of alternative splice variants in higher eukaryotes leading to enhanced functional annotation of the genes which might be useful in disease identification [46].

RNA sequencing is another important application of NGS where shotgun libraries are prepared from the total RNA or small RNA of a cell or organism and is sequenced which is useful for quantification of individual RNA species as well as their functional relevance is also revealed from the generated sequence information [47,48]. NGS analysis had identified differentially expressed miRNA target genes for acute myeloid leukemia [49]. In depth transcriptomics is essential for understanding the molecular composition of a cell to interpret its functional relevance in development and disease. Due to the widespread availability of faster and cheaper NGS technologies, it is possible to catalogue all transcripts and their expression levels under different developmental or physiological conditions.

DNA methylation studies have greatly benefitted from the advancements in NGS technologies as deep sequencing of bisulfite treated DNA can be done for large scale analysis of DNA methylation. NGS holds promise to substantially enhance the epigenomic research with a geno wide screening of methylation patterns among organisms which lead to transcriptional switch on/off of genes. ChIP-Seq technology helps us to capture DNA-protein interactions pulled down by chromatin immunoprecipitation and deep sequenced [50]. ChIP-Seq offers better coverage than microarray and reduces the dependency on probes. For example, using ChIP-seq on HT-NGS platform, the binding sites for a transcription factor (TF) and the human growth-associated binding protein (GABP alpha) were directly sequenced instead of being hybridized on a chip-array and unraveling the wide and intricate gene pathways regulated by PPARG gene.

NGS had accelerated the biomedical research by driving the search for novel biomarkers for cancer and other diseases by comparative analysis of normal and diseased cells and/or tissues [51]. Rapidity of NGS had enabled the rapid discovery of tumor-specific genomic rearrangements or somatic mutations responsible for cancer initiation and progression [52, 53] cost effective rapid NGS platforms had opened a new era of patient specific diagnosis of diseases so that each individual can be treated based on its own genomic composition advances in Table top sequencing machines had already made ‘personalized medicine’ a reality in few contexts but greater efforts are needed to make it more cost effective and accessible to larger population [54].

Third generation sequencing

There have been great advances in genomics due to the increased accessibility and continuous development of next generation sequencing technologies but the latest buzz has been the advent of third generation sequencing platforms. Though not fully developed, these have already started to complement the already existing NGS platforms. Primarily Pac Bio and Life Technologies had developed the single molecule template based sequencing platforms that break the sample at single molecule level with better read lengths due to the utilization of DNA polymerase and terminal phosphate labeled nucleotides. Third generation sequencing offers certain characteristic advantages over the second generation NGS platforms in terms of long read lengths, better resolution, easy sample preparation, significantly faster run times. Higher coverage of genomes at a faster speed gives the promise of using these platforms for clinical diagnostics and personalized medicine. The primary data analysis is much easier and straightforward in these single molecules based platforms, which may be useful to provide quick prognosis of a disease to the patient [55].

Presently, two major platforms are available for third generation sequencing which have slightly different chemistries and detection systems which will be briefly discussed in following subheadings.

Pacific Biosciences Single Molecule Real Time (SMRT)

This is a method for single molecule sequencing on a real-time basis commercialized by Pacific Biosciences, Inc., in 2010[56]. This combines nanotechnology with molecular biology and a sensitive technique to detect the fluorescence it utilizes the DNA polymerase enzyme to exploit the natural process of addition of nucleotides for the real time capturing of DNA synthesis. The library fragments are bound with the polymerase molecules and deposited at the bottom of the ZMWs. The fluorescently labelled nucleotides are added one at a time and incorporation releases the label. The instrument optics then detects these changes in each of the ZMW continuously thus enabling real time monitoring through a charged coupled diode (CCD) detection system. The ZMW chip is called a SMRT cell, which contains about 150,000 ZMWs. Depending on the library fragments and time of data collection one can obtain read lengths as large as 1-10 KBs [57] they have relatively shorter run times of 12-15mins nevertheless, the per run cost of the instrument is quite costly as compared to second generation sequencing platforms moreover, there may be a lag in the laser illumination in case of repetitive sequences which may decrease the accuracy in certain cases hence more powerful and durable CCDs are needed to increase the accuracy of this platform in future [57,58].

FRET

This is a real time single molecule based approach which harnesses fluorescence energy transfer and is commercialized by Life Technologies [59]. Similar to SMRT, it utilizes an engineered DNA polymerase that contains a donor fluorophore and fluorescently labeled nucleotides; each nucleotide is given one color coded label when a nucleotide is incorporated on DNA pol/template complex, it releases a FRET signal and the color-coded fluorophore lights up and is captured as a signal for that specific base addition [57, 60].

Data analysis: Bioinformatic tools

Data analysis is becoming more challenging everyday due to its rate limiting factor of storage, analysis and interpretation [61-63]. Bioinformatics is one of the major approach which has made this analysis easy for researchers. But still bioinformatic analysis itself is primary obstruction for researchers as everyone doesn't have access to a bioinformatic facility. These days' data analysis is upcoming and growing field which incorporates the several strategies and application of biological sciences with emerging advanced technologies. This is the combination of biology, information technology and computer science. Nowadays, technologies rely on statistical methods to analyze data. This approach is being used in several important fields including medical science, pharmacology and molecular biology.

Although each next generation sequencing platform has its unique pipeline and their companies provide their unique software for analysis to process the raw data and produce read files with high quality data but it is very important and helpful to have knowledge about some basic tools that can be used routinely for this advanced study. NCBI (The National Center for Biotechnology Information) is freely accessible to everyone that is very good platform for DNA-RNA sequence data, analysis and information. At NCBI, there are multiple direct links to access different databases, tools and utilities. It is one of the major resource for journals and books to get enormous information for each platform.

Some important tools for NGS data analysis are as follows:

- **SAM (Sequence Alignment/Map) Tools**: This tool allows you to manipulate alignments in SAM format.
- **FASTX Tool kit**: This is a very good collection of command line tools for short read fasta file pre-processing.
- **GATK (The Genome Analysis Toolkit)**: This tool is a structured software library which is very useful for NGS data as it includes a depth of coverage analyzers, a quality score recalibrator, a SNP/indel caller and a local realigner".
- **Varscan**: This tool is command line based that is used for detection of variant in massively parallel sequencing data.
- **CNVnator**: This tool is in use for CNV (copy number variation) discovery and genotyping from depth of read mapping.
- **Bowtie**: This tool is useful to aligns short DNA sequences (reads) to the human genome at a rate of over 25 million 35-bp reads per hour.

There are some packages in and market which contain all tools together and they are very easy to use like CLC Genomics Workbench, GenomeQuest and DNAnexus. These packages have its own algorithms based on application and they can be costly for researchers. To overcome this situation, there is one more platform called Galaxy. Galaxy has very easy web based system which uses linux command line tool. Once data is generated, that can be uploaded and mapped. Then it can be analyzed based on application without deep knowledge of command line.

CONCLUSIONS

NGS has become breakthrough in sequencing technologies by addressing complex biological questions. The reason behind this breakthrough is the accuracy of data information, cost effectiveness and ease of use with each new system. This technology has given a new direction in advancement of understanding of living systems and complex phenotypes. In this new era, scientists are rapidly moving towards personalized medicine and this field is getting revolutionary in a way that one can now diagnose the cancer and inherited diseases by sequencing the genome, one can provide personized treatment for patient's individual tumor or genetic mutation. One of the approach what scientists are using for precision diagnosis of infectious diseases called Metagenomics NGS. Therefore, it is easy for researchers and clinicians to understand the genetic variability as seen in every year in number of pertinent publications.

In conclusion, we are not very far to predict the near future of NGS. NGS has become an important technology and is in use as daily routine for researchers and clinicians. Our knowledge is increasing day by day in terms of human genome variability.

REFERENCES

1. Lam, H.Y. Clark, M.J. Chen, R. Chen, R. Natsoulis, G. O'Huallachain, M. Dewey, F.E. Habegger, L. Ashley, E.A. Gerstein, M.B. Butte, A.J. Ji, H.P. Snyder, M. Performance comparison of whole-genome sequencing platforms. Nat. Biotechnol. 2012, 30, 78–82.
2. Audo, I. Bujakowska, K.M. Leveillard, T. Mohand-Said, S. Lancelot, M. Germain, A. Antonio, A. Michiels, C. Saraiva, J. Letexier, M. Sahel, J. Bhattacharya, S.S. Zeitz, C. Development and application of a next-generation-sequencing (NGS) approach to detect known and novel gene defects underlying retinal diseases. Orphanet J. Rare Dis. 2012, 7, 8.
3. Dong, H. Wang, S. Exploring the cancer genome in the era of next-generation sequencing. Front. Med. 2012, 6(1), 48–55.
4. Shendure, J. et al., Advanced sequencing technologies: Methods and goals. Nature Rev. Genet. 2004, 5(5), 335–344.
5. Watson, J. Crick, F. Molecular structure of nucleic acids. Nature. 1953, 171, 709–756.
6. Zallen, D.T. Despite Franklin's work, Wilkins earned his Nobel, Nature. 2003, 425.
7. Sanger, F. Brownlee, G. Barrell, B. A two-dimensional fractionation procedure for radioactive nucleotides. J. Mol. Biol. 1965, 13, 373-IN4.

8. Brownlee, G. Sanger, F. Nucleotide sequences from the low molecular weight ribosomal RNA of Escherichia coli. J. Mol. Biol. 1967, 23, 337–353.
9. Cory, S. Marcker, K.A. Dube, S.K. Clark, B.F. Primary structure of a methionine transfer RNA from Escherichia coli. Nature. 1968, 220, 1039–1040.
10. Sanger, F. et al., Nucleotide sequence of bacteriophage phi X174 DNA. Nature. 1977, 265, 687–695.
11. Sanger, S. Nicklen, A.R.C. DNA sequencing with chain-terminating. Proc. Natl. Acad. Sci. 1977, 74, 5463–5467.
12. Smith, L.M. Fung, S. Hunkapiller, M.W. Hunkapiller, T.J. Hood, L.E. The synthesis of oligonucleotides containing an aliphatic amino group at the 5′ terminus synthesis of fluorescent DNA primers for use in DNA sequence analysis. Nucleic Acids Res. 1985, 13, 2399–2412.
13. Ansorge, W. Sproat, B.S. Stegemann, J. Schwager, C. A non-radioactive automated method for DNA sequence determination. J. Biochem. Biophys. Methods. 1986, 13, 315–323.
14. Prober, J.M. et al., DNA sequencing with rapid for system fluorescent chain terminating dideoxynuclcotides. Science. 1987, 238, 336–341.
15. Luckey, J. Drossman, H. High speed DNA sequencing by capillary electrophoresis. Nucleic Acids Res. 1990, 18, 4417–4421.
16. R.K. Saiki, S. Scharf, F. Faloona, K.B. Mullis, G.T. Horn, Enzymatic amplification of β-globin genomic sequences and restriction site analysis for diagnosis of sickle cell anemia. Science. 1985, 230, 1350–1354.
17. Saiki, R.K. et al., Primer-directed enzymatic amplification of DNA with a thermostable DNA polymerase. Science. 1988, 239, 487–491.
18. Cohen, S.N. Chang, A.C.Y. Boyert, H.W. Hellingt, R.B. Construction of biologically functional bacterial plasmids in vitro. Proc. Natl. Acad. Sci. U. S. A. 1973, 70, 3240–3244.
19. Venter, J.C. et al., The sequence of the human genome. Science. 2001, 291, 1304–1351.

20. Illumina®. Available online: http://www.illumina.com.
21. Ion-Torrent®. Available online: http://www.thermofisher.com.
22. PacBio®. Available online: http://www.pacb.com.
23. Roche®. Available online: http://www.roche.com.
24. Nanopore®. Available online: https://www.nanoporetech.com.
25. Margulies, M. Egholm, M. Altman, W.E. Attiya, S. Bader, J.S. Bemben, L.A. Berka, J. Braverman, M.S. Chen, Y.J. Chen, Z. et al., Genome sequencing in microfabricated high density picolitre reactors. Nature. 2005, 437, 376-380.
26. Gharizadeh, B. Ghaderi, M. and Nyren, P. Pyrosequencing technology for short DNA sequencing and whole genome sequencing. Technology. 2007, 47, 129-132.
27. Gharizadeh, B. Ohlin, A. Molling, P. Backman, A. Amini, B. Olcen, P. and Nyren, P. Multiple group-specific sequencing primers for reliable and rapid DNA sequencing. Mol. Cell Probes. 2003, 17, 203-210.
28. Simner, P. J. Khare, R. Wengenack, N. L. Rapidly Growing Mycobacteria. *In* Molecular Medical Microbiology: Second Edition. 2014, 3, 1679-1690.
29. Zhao, P. Bapat, B. The Role of Methylation-Specific PCR and Associated Techniques in Clinical Diagnostics. In Epigenetic Biomarkers and Diagnostics. 2016, 155–173
30. Morey, M. Fernández-Marmiesse, A. Castiñeiras. D, Fraga, J.M. Couce, M.L. Cocho, J.A. A glimpse into past, present, and future DNA sequencing. Mol. Genet. Metab. 2013, 110, 3–24.
31. Berger, M.F. Lawrence, M.S. Demichelis, F. Drier, Y. Cibulskis, K. Sivachenko, A.Y. Sboner, A. Esgueva, R. Pflueger, D. Sougnez, C. et al., The genomic complexity of primary human prostate cancer. Nature. 2011, 470, 214–220.
32. Kumar, A. White, T.A. MacKenzie, A.P. Clegg, N. Lee, C. Dumpit, R.F. Coleman, I. Ng, S.B. Salipante, S.J. Rieder, M.J. et al., Exome sequencing identifies a spectrum of mutation frequencies in advanced and lethal prostate cancers. Proc. Natl. Acad. Sci. USA. 2011, 108, 17087–17092.

33. Schweiger, M.R. Kerick, M. Timmermann, B. Isau, M. The power of NGS technologies to delineate the genome organization in cancer: From mutations to structural variations and epigenetic alterations. Cancer Metastasis Rev. 2011, 30, 199–210.
34. Kerick, M. Isau, M. Timmermann, B. Sultmann, H. Herwig, R. Krobitsch, S. Schaefer, G. Verdorfer, I. Bartsch, G. Klocker, H. et al., Targeted high throughput sequencing in clinical cancer settings: Formaldehyde fixed-paraffin embedded (FFPE) tumor tissues, input amount and tumor heterogeneity. BMC Med. Genomics. 2011, 4, 68.
35. Liu, L. et al., Comparison of Next-Generation Sequencing Systems. J. Biomed. Biotechnol. 2012.
36. Chiosea, S.I. Williams, L. Griffith, C.C. Thompson, L.D. Weinreb, I. Bauman, J.E. Luvison, A. Roy, S. Seethala, R.R. Nikiforova, M.N. Molecular characterization of apocrine salivary duct carcinoma. Am. J. Surg. Pathol.
37. Buzolin, A.L. Moreira, C.M. Sacramento, P.R. et al., Development and validation of a variant detection workflow for *BRCA1* and *BRCA2* genes and its clinical application based on the Ion Torrent technology. Hum. Genomics 2017, 11, 14.
38. Wright, C.F. Morelli, M.J. Thébaud, G. Knowles, N.J. Herzyk, P. Paton, D.J. Haydon, D.T. King, D.P. Beyond the consensus: dissecting within-host viral population diversity of foot-and-mouth disease virus using next-generation genome sequencing. J Virol. 2011, 85, 2266–2275.
39. Xi, R. Kim, T.M. Park, P.J. Detecting structural variations in the human genome using next generation sequencing. Brief Funct Genomics. 2010, 9, 405–415.
40. Turner, D.J. Keane, T.M. Sudbery, I. Adams, D.J. Next-generation sequencing of vertebrate experimental organisms. Mamm. Genome. 2009, 20, 327–338.
41. Wheeler, D.A. Srinivasan, M. Egholm, M. Shen, Y. Chen, L. McGuire, A. et al., The complete genome of an individual by massively parallel DNA sequencing. Nature. 2008, 452, 872–876.
42. McKernan, K.J. Peckham, H.E. Costa, G.L. McLaughlin, S.F. Fu, Y. et al., Sequence and structural variation in a human genome

uncovered by short-read, massively parallel ligation sequencing using two-base encoding. Genome Res. 2009, 19, 1527–1541.

43. Lind, C. Ferriola, D. Mackiewicz, K. Heron, S. Rogers, M. et al., Next-generation sequencing: the solution for high-resolution, unambiguous human leukocyte antigen typing. Hum. Immunol. 2010, 71, 1033–1042.

44. Burgess, D.J. Human Disease: Next-generation sequencing of the next generation. Nat Rev Genet. 2011, 12, 78–79.

45. Shendure, J. Mitra, R. D. Varma, C. Church, G. M. Advanced sequencing technologies: methods and goals. Nature reviews. Genetics. 2005, 5(5), 335.

46. Zhu, J. Shendure, J. Mitra, R. D. Church, G. M. Single molecule profiling of alternative pre-mRNA splicing. Science, 2003, 301, 836–838.

47. Ozsolak, F. Milos, P.M. RNA sequencing: advances, challenges and opportunities. Nat Rev. Genet. 2011, 12, 87–98.

48. Pareek, C. S. Smoczynski, R. Tretyn, A. Sequencing technologies and genome sequencing. J Appl Genet. 2011, 52(4), 413–435.

49. Ramsingh, G. Koboldt, D.C. Trissal, M. Chiappinelli, K.B. et al., Complete characterization of the microRNAome in a patient with acute myeloid leukemia. Blood. 2010, 116, 5316–5326.

50. Shendure, J., Ji, H. Next-generation DNA sequencing. Nature Biotechnology. 2008, 26, 1135 – 1145.

51. Pfeifer, G.P. Hainaut, P. Next-generation sequencing: emerging lessons on the origins of human cancer. Curr. Opin. Oncol. 2011, 23, 62–68.

52. Fang, X. Yoon, J.G. Li, L. Yu, W. Shao, J. Hua, D. Zheng, S. Hood, L. Goodlett, D.R. Foltz, G. Lin, B. The SOX2 response program in glioblastoma multiforme: an integrated ChIP-seq, expression microarray, and microRNA analysis. BMC Genomics. 2011, 12, 11.

53. Katsios, C. Ziogas, D.E. Liakakos, T. Zoras, O. Roukos, D.H. Translating Cancer Genomes Sequencing Revolution into Surgical Oncology Practice. J. Surg Res. 2010.

54. Meyerson, M. Gabriel, S. Getz, G. Advances in understanding

cancer genomes through second-generation sequencing. Nat Rev. Genet. 2010, 11, 685–696.

55. Thompson J.F. Patrice, M. M. The properties and applications of single-molecule DNA sequencing. Genome Biol. 2011, 12, 217.
56. Eid, J. Fehr, A. Gray, J. Luong, K. Lyle, J. Otto, G. Peluso, P. Rank, D. Baybayan, P. Bettman, B. et al., Real-time DNA sequencing from single polymerase molecules. Science. 2009, 323(5910), 133-138.
57. Munroe, D.J. Harris, T.J.R. Third-generation sequencing fireworks at Marco Island. Nat. Biotechnology. 2010, 28, 426–428.
58. M. J. Levene, J. Korlach, S. W. Turner, M. Foquet, H. G. Craighead, W. W. Webb Zero-Mode Waveguides for Single-Molecule Analysis at High Concentrations. Science. 2003, 299(5607), 682-686.
59. Vander Horn, P.B. Single Molecule Real-Time DNA Sequencing on the Surface of a Quantum-Dot Nanocrystal. J. Biomol. Tech. 2011, S9.
60. Blow, N. DNA sequencing: generation next-next. Nat. Methods, 2008, 5, 267-274.
61. Horner, D.S. Pavesi, G. Castrignano, T. et al., Bioinformatics approaches for genomics and post genomics applications of next generation sequencing. Brief Bioinform. 2010, 11, 181–97.
62. El-Metwally, S. Hamza, T. Zakaria, M. Helmy, M. Next-generation sequence assembly: four stages of data processing and computational challenges. PLoS Comput. Biol. 2013, 9, e1003345.
63. Land, M. Hauser, L. Jun, S.R. et al., Insights from 20 years of bacterial genome sequencing. Funct. Integr Genomics. 2015, 15, 141–61.

CHAPTER 2

TEMPLATE PREPARATION FOR THE PYROSEQUENCING METHOD

Prerna Pandey

Dr Prerna Pandey, a biotechnologist with several years of wet lab research experience, worked at International Center for Genetic Engineering and Biotechnology, New Delhi. Her field of research in PhD included isolation and molecular characterization of geminiviruses, genome sequencing, gene annotation, and gene silencing using the RNA interference technology. She has also worked at Transasia Biomedicals and Advance Enzyme Technologies as a scientist. Prerna has published papers in peer-reviewed journals, and has submitted a number of annotated Geminiviral genome sequences in the GenBank, including two novel ones. She has also completed her editing and proof-reading courses from Society for the promotion of Editors and Proof-readers, UK and now works as a freelance scientific editor and writer. When Prerna is not busy with her assignments, she loves to explore historical places.

INTRODUCTION

Pyrosequencing is an advanced technology to synthesize and monitor in real time a nucleotide sequence and involves enzymatic reactions (Ronaghi et al, 1998). This technology has an absolute requirement of

- a template primer hybrid

- enzymes: Klenow DNA polymerase, ATP sulfurylase, luciferase from firefly and potato apyrase. The role of apyrase is to degrade the nucleotides that fail to incorporate.
- enzyme substrates (adenosine 5′ phosphosulphate, also called APS and luciferin)
- nucleotides

Briefly, the polymerase facilitates the addition of a nucleotide complementary to the base on the template strand which results in release of pyrophosphate (PPi). The APS is converted to ATP by ATP sulfurylase using PPi. Luciferase/luciferin then metabolizes ATP to produce a flash of light, whose intensity is proportional to the number of incorporated bases. Addition of each nucleotide results in light peaks and sequence is determined from the order and relative magnitude of each light peak.

Since the enzymes used are heat labile, while the pyrosequencing reaction is typically carried out isothermally near 28°C, at such low temperature accessibility of the primer to the target region of DNA is hampered (Ronaghi, 2000). This can be avoided by sequencing single-stranded DNA templates, where, the primers do not have to compete with a tightly annealed complementary strand. ss DNA is obtained from double-stranded PCR products. One of the PCR primers are end-labelled with biotin which are pulled out using beads coated with streptavidin using a magnetic or vacuum equipment (Dunker, 2003) which leads to separation of labelled amplicon as well as PCR byproducts. A major disadvantage involved is that the multi-step nature of sample preparation process increases the risk of contaminating a laboratory with PCR products, and even minimal contamination represents a large fraction of the input sample.

TEMPLATE PREPARATION FOR THE PYROSEQUENCING METHOD

PCR products that are to be sequenced contain excess amounts of PCR primers, nucleotides and PPi, which have to be removed or degraded prior to sequencing by the Pyrosequencing method (Salk et al, 2006). These by-products can interfere in the course of DNA sequencing since the amplification primers can re-anneal to the amplified DNA, causing unspecific sequence signals, and the remaining nucleotides and PPi in the

PCR products should also be removed or degraded before the sequencing procedure is carried out to eliminate the risk for asynchronous extension and depletion of APS. Moreover, the salt in the PCR reaction inhibits the enzyme system slightly and should be removed or diluted. The use of an internal primer for DNA sequencing is preferential for higher specificity. Since there is a possibility or presence of unspecific amplification products in PCR, especially for genomic DNA, utilizing internal primers for sequencing can improve the sequence signal quality (Salk et al, 2006). The approaches that are currently used for template preparation are discussed briefly in this chapter.

Single-strand approach

As the name suggests, this approach is based on sequencing of single-stranded DNA template. In traditional Pyrosequencing method, solid-phase purification of the template, based on streptavidin biotin binding is utilized. One of the PCR primers is biotin-labeled at the 5'-terminus. The biotinylated PCR product binds to a streptavidin coated solid support, allowing the immobilization of the DNA fragment. This enables the immobilized DNA to become single-stranded by alkali treatment (high pH), and simultaneously, the interfering components are removed. The eluted nonbiotin-labeled strand can also be collected and used for sequencing (Gharizadeh, 2003).

After the separation step, a sequencing primer is annealed to the single-stranded DNA. There are two methods available for single-strand DNA separation

(i) streptavidin coated magnetic beads, using a magnet for purification, denaturation and elution, and (ii) streptavidin coated Sepharose beads, based on filter separation by a vacuum system.

An automated magnetic separation system is capable of performing the separation on 96-well plate format. The automated system can be programmed for different procedures, such as extra washing step to remove excess sequencing primers prior to DNA sequencing, and regeneration of magnetic beads for multiple use (Gharizadeh, 2003).

Double-strand approach

There are two different methods available for preparation of double-stranded template.

(i) In the first method (Nordstrom et al., 2000), nucleotides, PPi and amplification primers are removed enzymatically. The nucleotides and PPi are degraded by addition of apyrase and pyrophosphatase, and the amplification primers are removed by exonuclease I, which specifically degrades single-stranded DNA. The double-stranded DNA is then denatured by heat and the sequencing primer is annealed.

(ii) The second method is based on using blocking primers (Nordstrom et al., 2002). The principle behind this approach is to prevent the re-annealing of the present PCR primers to the amplified DNA. This could be carried out either by utilizing complementary blocking primers hybridizing to the PCR primers or using non-extendable primers, with the same sequence as the PCR primers, binding to the amplification sites. This later method is faster and the sequence results obtained show better signal quality.

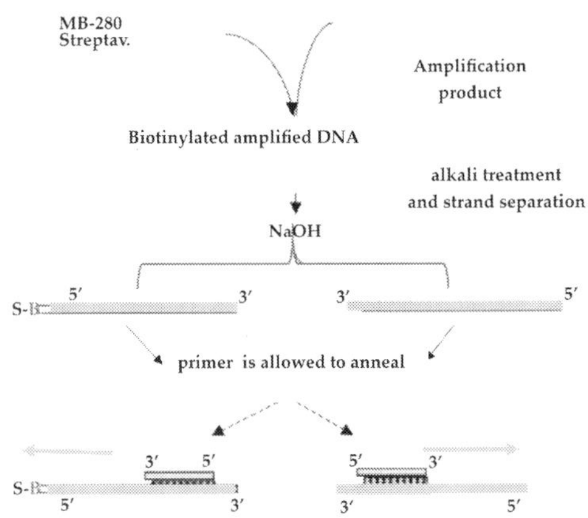

Figure 2.1: Single-stranded template preparation prior to the Pyrosequencing method utilizing steptavidin conjugated magnetic beads (S-B). The biotinylated amplified DNA binds to the magnetic beads and is separated under alkali treatment. The sequencing primers can be annealed to either strand for sequencing.

There are other protocols for template preparation for enhanced automation and speeding up the process of sample preparation (Salk et al., 2006). One of the method includes removal of PCR reaction components, which interfere with the sequencing process, with the help of enzymes and/or using blocking oligonucleotide (Nodstrom et al., 2000b). Yet another modification of the process is to generate single-stranded material by using only one 5′ phosphate-labeled primer for PCR, followed by digestion with lambda exonuclease (Diggle and Clarke, 2003). Though this process yields a single-stranded template it has a drawback that it involves lengthy workup and the PCR byproducts need to be removed by solid-phase filtration, adding to the instrumentation.

LINER-AFTER-THE-EXPONENTIAL (LATE]-PCR METHOD

Although Pyrosequencing of double-stranded (ds) DNA templates has been performed, the technique works best if single-stranded (ss) templates are used (Salk et al., 2006; Agah et al., 2004). To obtain ss DNA, a PCR is performed with a biotinylated primer which is followed by alkaline denaturation of the ds DNA. Subsequently, the DNA strand with incorporated biotinylated primer binds selectively to streptavidin-coated surfaces such as magnetic beads and chromatographic resins. Liner-After-The-Exponential [LATE]-PCR, a highly robust asymmetric PCR method has been devised to get ssDNA for pyrosequencing. This method makes use of unequal concentrations of primers to generate large amounts of single-stranded DNA (ssDNA) along with a small fixed amount of double-stranded DNA (dsDNA) (Salk et al., 2006; Sanchez et al., 2004; Pierce et al., 2003). Thus, the ss DNA templates are generated and the remaining pyrophosphate and excess amounts of unbiotinylated primer and dNTPs from PCR are removed to eliminate any interference with the sequencing process. This method takes care of interfering PCR reaction components by using enzymes to break them down. Since this process yields ss-DNA, the process is simplified as there is no need for strand isolation and it can be directly used for pyrosequencing. This also helps in minimizing the contamination risk to the entire process as it eliminates the need for time-consuming solid-phase workup and can be accomplished in a single reaction vessel without expensive biotinylated primers, streptavidin beads or purification equipment. Moreover, it obviates the requirement of primer removal by exonucleases prior to pyrosequencing (Salk et al., 2006; Nodstrom et al, 2000b) and blockage of primers or primer binding sites with 3′ phosphorylated complementary sequences (Nodstrom et al, 2002). Thus, it is a fast technique as compared to others.

Development of fully automated one tube LATE-PCR can be facilitated by using microfluidic or mesofluidic device where multiple assays can be done parallelly on a single chip or wafer. Pyrosequencing chemistry has already been applied on bead-immobilized DNA in microfabricated chips (Salk et al., 2006; Russom et al., 2003). Such processes find a great application where low cost, region-specific sequencing is needed, for example in diagnostic, forensic and bio-detection devices, which can be made in an automated and cost-effective manner.

VECTOR LIGATION AND UNIVERSAL BIOTINYLATED PRIMERS SINGLE-STRANDED DNA TEMPLATES

Another approach used to obtain ssDNA was used by Pacey-Miller and Henry (2003) who developed a PCR protocol in which a "11-base GC tag oligonucleotide" is used in a first PCR; the GC tag is targeted in a second PCR employing a GC tag-specific biotinylated primer (Groth et al., 2006). However, this approach might not be applicable for Pyrosequencing analyses in GC-rich regions or CpG islands characteristic of mammalian genomes. Mis-priming of the GC tag could occur and would require the use of tags that do not interfere with the respective sequence context. To overcome the above-mentioned shortcomings a method was designed by Colella and coworkers (Colella et al., 2003). Their protocol involves the use of three primers in one PCR, namely template-specific forward and reverse primers and one universal biotinylated primer. The template-specific reverse primer harbors at its 5' end 20 nucleotides (nt) that comprise the sequence of a universal biotinylated primer. This tag differs from the tag used by Pacey-Miller and Henry (2003) in length (20 vs. 11 nt) and GC content (55 vs. 100%) and, therefore, can be used to analyze GC-rich regions. However, laborious optimization of the PCR protocol is needed; for example, ratios of all primers need to be adjusted.

Guo and Milewicz (2003) devised a similar approach in which, PCR-amplified DNA is labeled with biotin for Pyrosequencing and with FAM and HEX for microsatellite genotyping. Recently, Aydin and coworkers (Aydin et al., 2006) introduced a universal biotinylated primer method similar to the approach of Pacey-Miller and Henry (2003). It involves a PCR using two target-specific primers and two different universal primers, one of which is biotinylated. Each target-specific primer harbors a sequence corresponding to one of the universal primers at its 5' end. In addition to requirements for template preparation, it is necessary for certain ss templates to prevent folding back of their endings; this entails self-priming and interferes with Pyrosequencing (Groth et al., 2006; Utting et al., 2004).

Although there are several methods that greatly reduce the number of biotinylated primers necessary for preparation of ss templates, they always are applicable to distinct analyses only. Thus, the development of a simple methodology would represent a cost- and time-efficient improvement for Pyrosequencing if many different targets need to be analyzed.

Later, a universal method for sequencing of DNA fragments was developed by Anderson and coworkers, in which the DNA fragments were cloned into an ss phage vector, and ss template preparation was optimized for sequence determination by Pyrosequencing. This method involves two steps:

- First, a PCR-amplified DNA target is ligated into a linearized DNA vector, and
- Second, this recombinant DNA serves as a template for a PCR using one vector-specific biotinylated primer combined with a target-specific (unlabeled) primer. this method was applied to genotype two known single-nucleotide polymorphisms (SNPs) located in the human CARD15 and A2M genes and for characterization of four multisite variations (MSVs) (Fredman et al., 2004) in the human DEFB104 gene, which maps to a segmentally duplicated region characterized by a high polymorphism rate (Groth et al., 2006).

International HapMap Project (2003) and other large SNP typing projects, that have provided invaluable data widely used for evaluation of genetic differences associated with individual phenotypes and disease gene mapping (Peacock and Whiteley, 2005). However, even though millions of SNPs are deposited in numerous public databases and accessible to individual researchers, there is an ongoing need for efficient high-throughput SNP typing. Genotyping or allele frequency determination by Pyrosequencing requires ss templates. Universal primers have been used for convenient Sanger sequencing of ss recombinant DNA fragments, as well as Pyrosequencing (Anderson et al., 1980). The method is easy, fast, and straightforward, and it can be adapted to other vector systems and DNA joining techniques. The advantage of this approach, compared with published methods using universal primers (Groth et al., 2006), is that

- primers and PCR protocols to amplify the target sequence can be defined and optimized regardless of the Pyrosequencing reaction. Because one target-specific oligonucleotide is used in the second PCR and also as the sequencing primer, and because its annealing site is adjacent to the sequence of interest, the 3' end of the ss template is locked and self-priming of the template is prevented (Utting et al., 2004).
- Another advantage of this approach is the nondirectional insertion of PCR-amplified target into vector, allowing Pyrosequencing of both strands with two target-specific sequencing primers and only one universal biotinylated primer. is the technique can be used where multiplexed target-specific PCRs are to be performed and more than one amplicon is to be inserted into a plasmid vector in a single reaction for high throughput genotyping platform. Here, in the second PCR, several target regions can be amplified from the pool of recombinant molecules carrying different inserts by using respective target-specific primers and only one biotinylated primer.

This procedure of labelling ds DNA by PCR with a biotin tag facilitating downstream processing by means of biotin–streptavidin interactions for Pyrosequencing is simple and universal. Moreover, it can be adapted for labeling of ds DNA with other markers, such as FAM, HEX, and NED, useful for mutation screening by ss conformation polymorphism analysis (Aydin et al., 2004) or for microsatellite genotyping (Guo et al., 2003).

METHOD FOR ONE-STEP PREPARATION OF DOUBLE-STRANDED DNA TEMPLATE

Direct use of double-stranded DNA (dsDNA), instead of single-stranded DNA (ssDNA), has simplified the template preparation step prior to performing an analysis using Pyrosequencing (Nordstrom et al., 2002). Purity of sequencing of ds DNA obtained as a PCR product is of great concern. Excess amounts of primers and nucleotides after the amplification step must be removed prior to sequencing. Interfering components of the PCR mixture can be removed by filtration, chromatography or gel purification methods. Enzymatic methods have also been employed for the removal of interfering components (Nordstrom et al., 2002). While apyrase and inorganic pyrophosphatase (PPase) very efficiently degrade nucleotides and PPi, exonuclease I degrades the excess PCR primers. However, the use of exonuclease I in combination with apyrase and PPase for sample preparation require two steps prior to use in this case. Furthermore, the low Kcat of exonuclease I determines the limit for how fast the procedure can be performed. Here, we describe a new strategy for preparation of dsDNA (Li et al., 1991). In the new strategy, problems with nonspecific priming due to primer – dimers and loops are avoided. In addition, by exclusion of the exonuclease I treatment, the time and steps necessary for efficient template preparation is reduced (Nodstrom et al., 2002).

The strategy of primer blocking

In the new template preparation protocol, PCR-amplified dsDNA is incubated with sequence primer, blocking oligonucleotides, apyrase and PPase at 35 ° C for 4 min. The sample is then heated at 100 °C for 2 min, and thereafter, rapidly cooled at 20 °C. In the first incubation step, deoxynucleotides and PPi, that remain unutilized from the PCR, are hydrolyzed. The heating step is necessary for inactivation of the enzymes and for efficient melting of the dsDNA. Fast cooling is necessary for efficient hybridization of sequencing primer and blocking oligonucleotides. This methodology involves the use of two different types of blocking oligonucleotides is very helpful in preventing the remaining PCR primers to reanneal to the template, which is of great concern:

- oligonucleotides complementary to the primers used for the PCR and
- 3'-end modified oligonucleotides with the same sequence as the primers used for the PCR. The 3'V-ends of the oligonucleotides can be modified by addition of a few mismatch basesor by a blocking group (e.g. a phosphate or an amino group)

(Nodstrom et al., 2002).

CASE STUDIES

Sequencing on dsDNA using oligonucleotides complementary to the PCR primers

In the first variant of the new approach, remaining PCR primers are prevented from reannealing by using complementary oligonucleotides. During the template preparation step, remaining PCR primers are thought to preferably hybridize to an excess of added oligonucleotides complementary to the PCR primers. False signals were observed due to priming by one of the remaining PCR primers. In this case correct sequence could be determined by the addition of an oligonucleotide (B) complementary to the second PCR primer (Nodstrom et al, 2002; Hanke and Wink, 1994). However, if instead of the addition of a complementary oligonucleotide, an excess of the second PCR primer was added to the PCR product, the false signals increased, clearly confirming that the false signals were due to remaining PCR primers. Also, in this latter case, the false signals could be abolished by addition of complementary oligonucleotides (Nodstrom et al., 2002; Nodstrom et al., 2000a,b).

In the example described above, one of the PCR primers was used as sequencing primer. However, if an internal sequencing primer is used instead, two different blocking oligonucleotides must be added during the template preparation procedure. In the presence of blocking oligonucleotides, complementary to both PCR primers, accurate data from the whole (35 bases) region can be generated (Nodstrom et al., 2002).

Sequencing on dsDNA using 3'-end modified PCR primers

In the second approach, 3'-end modified oligonucleotides, with the same sequence as the primers used for PCR, are used as blocking oligonucleotides. Initially, to investigate the effectiveness of the blocking strategy, two different 3'-end modified oligonucleotides were tested—one terminated at the 3'-end by a phosphate group and one extended by five mismatch bases at the 3'-end (Nodstrom et al., 2002). When the test was performed on a 222-base-long dsDNA template, false signals were observed due to priming by the remaining PCR primers and unspecific, self-priming of the free 3'-end of one of the template strands. In the presence of the blocking oligonucleotides, accurate data from the whole (31 bases) sequenced region could be generated. The same result was obtained when both blocking oligonucleotides were 3'-end terminated. It is worth noting that 2.5 Ag SSB was present in the Pyrosequencing reaction mixture, although a similar result (with slightly higher backgrounds) was obtained in the absence of SSB. There are some advantages of using 3'-end blocked oligonucleotides, with the same sequence as the PCR primers, compared to oligonucleotides complementary to the PCR primers. For example, nonspecific hybridization and priming of free 3'-ends of the template strands can be avoided. In addition, when complementary PCR primers are used as blocking oligonucleotides, the quality of the oligonucleotides used should be considered, e.g. full-length, correct sequence and termination efficiency (termination efficiency is also an important factor when 3'-end blocked oligonucleotides are used) (Nodstrom et al., 2002).

Nonspecific signals and signal-to-noise ratio

Different types of unspecific 3'-end hybridization, i.e. mispriming, loops and dimers, can lower the quality of sequencing performance by decreasing the signal-to-noise ratio. Mispriming can occur due to unspecific hybridization of the different oligonucleotides (sequence primer, blocking oligonucleotides, remaining PCR primers) to the template. Loops and dimers can form from oligonucleotides and DNA strands. To overcome obstacles with nonspecific hybridization of free 3'-ends during Pyrosequencing analysis, primer design, blocking oligonucleotides and/or utilization of SSB is crucial. For instance, when one of the strands (the strand that was not sequenced) of the template was bound to paramagnetic beads, sequence signals were obtained in the absence of sequence primer due to nonspecific self-priming of the free 3'-end of the ssDNA. In addition, very strong signals observed in the presence of a primer-dimer (Hanke and Wink, 1994) disappeared if SSB was added to the sequencing reaction mixture. The signals also decreased in the presence of complementary oligonucleotides, but not to the same extent.

Advantages with the new strategy includes

- it is a one-step, faster and simpler template preparation process which is
- there is no need for exonuclease I treatment
- there is less problem with unspecific priming from loop structures and dimers.

(Nodstrom et al., 2002).

Preparation of Template by PCR for Massively Parallel Pyrosequencing

Massively parallel pyrosequencing involves immobilizing DNA fragments on beads for genome survey sequencing and transcriptome analysis of a variety of genomes. However, preparation of sufficiently high-quality template for pyrosequencing has been a challenge in analysis of nucleic acids sequences from tissues or cell types available in limited quantities.

The use of a biotinylated primer for polymerase chain reaction amplification allows removal of excess primer and poly(A) tract fragments from the sequencing templates, providing much higher yields of useful sequence information from pyrosequencing of amplified templates. This minimizes the amount of tissue sample required for deep sequencing analysis (Whetten et al., 2009), and allows deep sequencing analysis of nucleic acids isolated from very small tissue samples.

Whetten and coworkers obtained over 220 Mb of transcript DNA sequences from *Abies fraseri* (Pursh) Poir., a conifer species native to the southern Appalachian Mountains of eastern North America (Whetten et al., 2009). The method proved its utility on comparison of the resulting assembled putative transcripts with similar data obtained from other conifers by other sequencing methods. They observed that the use of a biotinylated primer in the final stage of cDNA preparation prior to nebulization and ligation of sequencing adaptors resulted in a much lower incidence of the DNA sequence read contamination by primer sequences. The presence of biotin on the 5' ends of PCR-amplified cDNAs and primers allowed trapping of excess PCR primer and cDNA fragments containing the biotinylated ends of amplicons on the streptavidin-coated beads used in preparation of template for emulsion-PCR (Margulies et al., 2005). This trapping probably removed most of the polyA tail fragments from the nebulized cDNA preparation. Thus, this selective removal of end fragments from the cDNA preparation did not significantly affect the distribution of sequence reads or the ability to identify proteins homologues by using assembled sequence contigs from Blastx queries of a protein database.

SUMMARY

Pyrosequencing has great application in preliminary investigations of species that have not yet been the subject of significant genomic research. Such studies and genomic surveys leads to sequences and catalogs of expressed genes and prove to be of great use in forming a link in the biology of less intensively studied species to that of more intensively studied model organisms.

In this chapter, we discussed different types of integrated, efficient and cost-effective system for template preparation strategies for pyrosequencing. The described procedures are simple, requiring little handling time, especially for high-throughput sequence-based analyses.

REFERENCES

1. Agah A, Aghajan M, Mashayekhi F, Amini S, Davis RW, Plummer JD, Ronaghi M, Griffin PB. A multi-enzyme model for pyrosequencing. Nucleic Acids Res. 2004; 32: e166.
2. Anderson S, Gait MJ, Mayol L, Young IG. A short primer for sequencing DNA cloned in the single-stranded phage vector M13mp2. Nucleic Acids Res. 1980; 8: 1731–1743.

3. Andersson H, van der Wijngaart W, Stemme G. Micromachined filter-chamber array with passive valves for biochemical assays on beads. Electrophoresis. 2001; 22: 249–257.
4. Aydin A, Luft FC, Bahring S. Validation of Xuorescence-labeled artiWcial nonhuman sequences for single-strand conformation polymorphism mutation detection in familial hypercholesterolemia, Anal Biochem. 2004; 324: 16–21.
5. Aydin A, Toliat MR, Bahring S, Becker C, Nurnberg P. New universal primers facilitate Pyrosequencing. Electrophoresis. 2006; 27: 394–397.
6. Colella S, Shen L, Baggerly KA, Issa JP, Krahe R. Sensitive and quantitative universal Pyrosequencing methylation analysis of CpG sites. BioTechniques. 2003; 35: 146–150.
7. Diggle MA, Clarke SC. A novel method for preparing singlestranded DNA for pyrosequencing. Mol Biotechnol. 2003; 24: 221– 224.
8. Dunker J, Larsson U, Petersson D, Forsell J, Schiller AL, Alderborn A, Berg LM. Parallel DNA template preparation using a vacuum filtration sample transfer device. Biotechniques. 2003; 34: 862– 866.
9. Ehn M, Nourizad N, Bergstrom K, Ahmadian A, Nyrén P, Lundeberg J, Hober S. Toward pyrosequencing on surface-attached genetic material by use of DNA-binding luciferase fusion proteins. Anal Biochem. 2004; 329: 11–20.
10. Fredman D, White SJ, Potter S, Eichler EE, Den Dunnen JT, Brookes AJ. Complex SNP-related sequence variation in segmental genome duplications. Nat Genet. 2004; 36: 861–866.
11. Gharizadeh B. Method development and applications of Pyrosequencing technology. Doctoral dissertation from the Department of Biotechnology, Royal Institute of Technology, Stockholm, Sweden, 2003. ISBN 91-7283-609-1.
12. Groth M, Huse K, Reichwald K, Taudien S, Hampe J, Rosenstiel P, Birkenmeier G, Schreiber S, Platzer M. Method for preparing single-stranded DNA templates for Pyrosequencing using vector ligation and universal biotinylated primers. Anal Biochem. 2006; 356: 194–201.
13. Guo DC, Milewicz DM. Methodology for using a universal primer to label ampliWed DNA segments for molecular analysis. Biotechnol.

Lett. 25 2003; 25L 2079–2083.

14. Hanke M, Wink M. Direct DNA sequencing of PCR-amplified vector inserts following enzymatic degradation of primer and dNTPs. BioTechniques 1994; 17: 858 – 860.
15. Li HH, Cui XF, Arnheim N. Eliminating primers from completed polymerase chain reactions with exonuclease VII. Nucleic Acids Res. 1991; 19: 3139 – 3141.
16. Margulies M, Egholm M, Altman WE, et al. Genome sequencing in microfabricated high-density picolitre reactors. Nature. 2005; 437: 376–380.
17. Nordström T, Ronaghi M, Forsberg L, de Faire U, Morgenstern R, Nyrén P. Direct analysis of single-nucleotide polymorphism on double-stranded DNA by pyrosequencing. Biotechnol Appl Biochem. 2000b; 31: 107–112.
18. Nordstrorm T, Anders Alderborn A , Nyre´n P. Method for one-step preparation of double-stranded DNA template applicable for use with Pyrosequencingk technology. J Biochem Biophys Methods 52 (2002) 71 – 82.
19. Nordstrorm T, Nourizad K, Ronaghi M, Nyre´n P. Method enabling pyrosequencing on double-stranded DNA. Anal Biochem. 2000a; 282: 186 – 93.
20. Nusbaum C, Mikkelsen TS, Zody MC, Asakawa S, Taudien S, Garber M, et al., DNA sequence and analysis of human chromosome 8. Nature. 439 (2006) 331–335.
21. Pacey-Miller T, Henry R. Single-nucleotide polymorphism detection in plants using a single-stranded Pyrosequencing protocol with a universal biotinylated primer. Anal Biochem. 2003; 317: 166–170.
22. Peacock E, Whiteley P. Perlegen Sciences, Inc., Pharmacogenomics. 2005; 6: 439–442.
23. Pierce KE, Rice JE, Sanchez JA, Wangh LJ. Detection of cystic fibrosis alleles from single cells using molecular beacons and a novel method of asymmetric real-time PCR. Mol Hum Reprod. 2003; 9: 815–820.
24. Ronaghi M, Uhlen M, Nyrén P. A sequencing method based on realtime pyrophosphate. Science. 1998; 281: 363–365.

25. Ronaghi M. Improved performance of pyrosequencing using singlestranded DNA-binding protein. Anal Biochem. 2000; 286: 282–288.
26. Russom A, Tooke N, Andersson H, Stemme G. Single nucleotide polymorphism analysis by allele-specific primer extension with real-time bioluminescence detection in a microfluidic device. J Chromatogr A. 2003; 1014: 37–45.
27. Salk JJ, Sanchez JA, Pierce KE, Rice JE, Soares KC, Wangh LJ. Direct Amplification of Single-Stranded DNA for Pyrosequencing using Linear-After-The-Exponential (LATE)-PCR. Analytical biochemistry. 2006; 353(1): 124-132.
28. Sanchez JA, Pierce KE, Rice JE, Wangh LJ. Linear-After-TheExponential (LATE)-PCR: an advanced method of asymmetric PCR and its uses in quantitative real-time analysis. Proc Nat Acad Sci USA. 2004; 101: 1933–1938.
29. Taudien S, Galgoczy P, Huse K, Reichwald K, Schilhabel M, Szafranski K, et al. Polymorphic segmental duplications at 8p23.1 challenge the determination of individual defensin gene repertoires and the assembly of a contiguous human reference sequence. BMC Genomics. 2004; 5: 92.
30. The International HapMap Project, Nature 426 (2003) 789–796. [23] D.A. Hinds, L.L. Stuve, G.B. Nilsen, E. Halperin, E. Eskin, D.G. Ballinger, K.A. Frazer, D.R. Cox, Whole-genome patterns of common DNA variation in three human populations. Science. 2005; 307: 1072– 1079.
31. Utting M, Hampe J, Platzer M, Huse K. Locking of 3 ends of singlestranded DNA templates for improved Pyrosequencing performance. BioTechniques. 2004; 37: 66–73.
32. Whetten RW, Sofía VA, Frampton J. Polymerase Chain Reaction Preparation of Template for Massively Parallel Pyrosequencing. J Biomol Tech. 2009; 20(2): 128-134.

CHAPTER 3

METHODOLOGICAL IMPROVEMENTS IN PYROSEQUENCING

Prerna Pandey

Dr Prerna Pandey, a biotechnologist with several years of wet lab research experience, worked at International Center for Genetic Engineering and Biotechnology, New Delhi. Her field of research in PhD included isolation and molecular characterization of geminiviruses, genome sequencing, gene annotation, and gene silencing using the RNA interference technology. She has also worked at Transasia Biomedicals and Advance Enzyme Technologies as a scientist. Prerna has published papers in peer-reviewed journals, and has submitted a number of annotated Geminiviral genome sequences in the GenBank, including two novel ones. She has also completed her editing and proof-reading courses from Society for the promotion of Editors and Proof-readers, UK and now works as a freelance scientific editor and writer. When Prerna is not busy with her assignments, she loves to explore historical places.

INTRODUCTION

The study of organisms nowadays entails the use of the composition of the genetic material in such a manner that the methodology applied is rapid, sensitive as well as applicable with a high throughput. The technique has seen many changes right from its initial method using chain termination using dideoxy nucleotides. The quest for other technologies spurned the development of other sequencing technologies such as hybridization or ligation and pyrosequencing (Mostafa Ronaghi, 2001).

The technique of pyrosequencing is based on bioluminescence for the study of several aspects such as single nucleotide polymorphisms (SNPs), typing of bacteria, fungi and virus. Improvements in technology have extended in the detection of mutations, methylation patterns, analysis of clones as well as secondary structure studies (Gharizadeh *et al*, 2006).

Certain challenges presented in the pyrosequencing technique include the development of single stranded samples as well as the design of primers to the administration of samples as well as reagents. The use of single-stranded DNA binding protein (SSB) has been applied to reduce the extent of primers and templates. The technique, however, faces limitations such as its limitation at a resolution of secondary structures apart from primer hybridizations (Gharizadeh *et al*, 2006).

The initial years of its advent, the technique of pyrosequencing was limited to 20 nucleotides from the primer as by products accumulated in the system led to an increase in noise. With several improvements in the methodology: the technique has been enhanced for longer stretches of DNA as well as homopolymer regions (Gharizadeh, 2003).

IMPROVEMENT OF READ LENGTH USING MODIFIED BASES

The technique of Pyrosequencing offers advantages such as the utilization of primers or nucleotides that are non-radioactive apart from its accurate and flexible technology. However, the activity of apyrase enzyme to degrade nucleotides decreased with the reaction that led to the technique being used only for short reads. Initially as the technique was applied, the use of dATP for luciferase gave rise to false positive results. In order to enhance the signal-to-noise ratio, dATPaS was used to replace dATP (Ronaghi *et al*., 1996).

Further studies using showed that the use of dATPαS resulted in a continued drop in the activity of apyrase apart from aberrant use of adenine nucleotide in regions that were homopolymers of Thymine. This dATPαS applied was a combination of Sp and Rp: two isomers. Subsequently, studies were performed for ATP-luciferase assays of different levels of pure Sp-isomer as well as at various Sp/Rp ratios. It was reported that the Klenow polymerase did not utilize the Rp isomer as its substrate. The successful incorporation of adenosine nucleotides in homoloymeric T regions involved the use of greater ratios of mixture Sp/Rp. Efficient incorporation of A was achieved by using half the level of pure Sp-isomer. The use of the pure Sp-isomer also resulted in an increase of the lengths of the reads. It was reported that inhibition of apyrase activity was not observed when alkaline phosphatase was used in the system suggestive of the products of degradation such as nucleoside diphosphates interfering with the activity of apyrase. As compared to regular nucleotides, the formation of alpha-S Nucleotide diphosphates resulted in a greater inhibition of the apyrase with decrease in activity as compared to the regular nucleoside diphosphates.

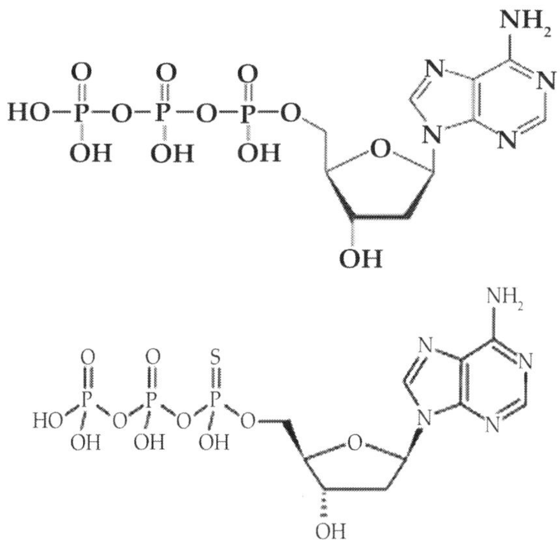

Figure 3.1: Structures of dATP (left) and dATPαS (right).

The utilization of pure Sp-dATPαS on different templates revealed that though the sequencing reaction depends on the template used; higher reads of more than 100 bases were achieved (Gharizadeh *et al*, 2002).

INCREASING READ LENGTHS USING SSB

E. coli Single-stranded DNA binding protein (SSB) confers protection to single stranded DNA by preventing its degradation or secondary structures (Meyer and Laine, 1990). A study in 2002 evaluated the use of

SSB in a pyrosequencing technology using PCR products from a cDNA library as templates (Agaton C, *et al*, 2002).

The results of addition of SSB are summarized as follows:

- The length of reads of the PCR products showed an increase with the application of SSB as compared to controls.
- With an increase in length of the PCR product, the sequence quality also showed a dip in the sequencing due to background noise that has been attributed to primer dimmers. The addition of SSB caused a reduction in the noise.
- Due to insufficient activity of apyrase, the nucleotides are not degraded completely between the cyclic addition of reagents that lead to plus shift mutations. The addition of SSB caused a reduction in the occurrence of such plus mutations.
- In homoplolymeric stretches, an insufficient activity of Klenow polymerase, which in turn would cause a lack of extension leading to a minus shift mutation. The adition of SSB caused a reduction in such shift mutations, its mode of action being a breakdown of secondary structures formed in such homopolymeric regions.
- The limiting factors were decreased by the use of SSB when short or moderate sized templates resulting in longer reads.
- For longer templates, the SSB abolished background sequence leading to larger reads. More than 25 bases could be assessed despite other factors that limited such long clones.

Taking together all these observations, these results show that SSB can be applied for a quick as well as elementary application in pyrosequencing for improving the background as well as read lengths.

(Reviewed by Afshin Ahmadian *et al*, 2006).

The application of SSB for the sequencing of amplicons of Human papillomaviruses (HPV): (HPV-72 and HPV 31.40/73) using a pool of 7 primers resulted in improved results of sequencing when SSB was utilized (Nordstrom *et al*, 2001).

IMPROVEMENT IN SEQUENCING OF HOMOPLOYMERIC T REGIONS:

For efficient use in sequencing, the enzyme used as polymerase should use modified nucleotides efficiently irrespective of the nature of the templates used. The enzyme should also lack 3'-5' nuclease activity to prevent the hydrolysis of primer that has annealed the templates. An added advantage would be the rejection of loops or primer dimmers as substrates for extension. As Klenow polymerase displayed these properties, it was used for pyrosequencing.

The sequence read length was increased by the use of pure Sp-dATPαS however; homopolymeric T-regions presented an issue for synchronous sequencing. The use of the modified nucleotides with alpha thio groups resulted in a low efficiency of catalysis by the Klenow polymerase. This in turn resulted in inefficient use of the dATPαS in homoploymeric regions (greater than 4-5 residues) of Thymine nucleotides.

With a purpose of improvement in the incorporation of nucleotides, Sequenase™: a T7 DNA polymerase that lacks exonuclease activity was tested on templates that contained several sequences of polymeric T residues using a parallel pyrosequencing technique.

While, the Klenow polymerase displayed lack of synchronous results when the numbers of T residues were more than 4, the Sequenase™ incorporated Adenosine with higher efficiency. The incorporation of other nucleotides was found to be equivalent for both the Sequenase™ as well as Klenow polymerase.

The use of Sequenase™ conferred added merits that included:

- Greater reads for lengths of homopolymeric T regions along with efficient incorporation of modified residues.
- The enzyme also could differentiate between the primer-template binding as well as primer dimmers.
- It had a better ability to recognize mismatches at 3' end than Klenow polymerase.
- The use of loop structures as primers was not favoured by the T7 enzyme as compared to Klenow polymerase (Gharizadeh, 2003).

PREPROGRAMMED SEQUENCING STRATEGY

The platform of pyrosequencing presents an important as well as distinct aspect of addition of nucleotides in a pre-programmed format. In a *de novo* sequencing, nucleotides are added in a cyclic format that leads to a longer duration of the reaction as well as accruing of samples. The pre-programmed system involves the addition of nucleotides in a pre-determined format in the reaction chamber.

The technique of pre-programmed pyrosequencing was used for checking clones for the presence of artefacts. Additionally, the verification of a proper insert and frame of the gene of interest in the vector is vital as well as analysis of any unwanted mutations. The technique of pyrosequencing for quality control yielded success with the detections of mutations: insertions or deletions easily visualized in the program in the form of peaks. Even mutations at single bases, in a stretch of repeated bases could be visualized as shifts in the heights of the signal peaks. The clones of apyrase gene were checked for deletion or insertion or mutations. A clone with a base deleted at one position was identified as a missing peak; as well as a clone with a deletion of 195 nucleotides could also be identified with the use of the pre-determined format.

The pre-programmed dispensation was reported as advantageous in the following aspects (Nourizad *et al*, 2003):

- The longer fragments could be read by the use of primers that bind to various regions of the template with a gap of 100-150 bp.
- The artefacts could be efficiently detected
- The clones or mutation could be assessed in a rapid fashion.

MULTIPLE SEQUENCING PRIMERS

In typing studies of microbes: universal or degenerate primers are used that recognize conserved sequences in order to recognize a broad range of organisms. The sequences that vary within a gene or the variable sequences are used for identification and detection. The use of such primers can have the potential to amplify other unintentional targets of the genome that can lead to non-specifics. Due to several species present in an amplicon or this non-specific binding: the field of sequencing faces a challenge (Gharizadeh *et al*, 2001).

For overcoming the problem of data that is complex to analyze to to a slew of species in a sample: pattern recognition system can be employed. The technique involves aligning the results of analysis of sequences to identify the characteristic feature in each type. An equivalent number of patterns are expected on the basis of the numbers of oligonucleotide primers used in a system with one pattern applied for a primer. The process of such pattern recognition could rely on the choice of the primers as well as the choice of an appropriate downstream sequence. This can entail the use of a software for Pattern recognition to facilitate pattern identification with ease. Pyrosequencing thus offers the advantages in

- The design of primers with flexibility to target specific regions.
- The design of primers is not rigid as sequencing is initiated at the first base downstream of the primer.
- The advantages of the pre-programmed format of addition of bases.
- As the design of the primers can select sequence reads that are short the process would possess more accuracy as well as rapid and economical.
- With reduced costs, parallel sequencing of several samples can be performed each with varying pools of primers to target several types of targets in a sample.

A 2001 team designed 7 primers that were specific for 7 important genotypes of Human papilloma viruses (HPV) that show multiple species in a clinical sample. The application of the pool of primers to samples enabled the sequencing of the genotypes with specificity. The use of sequence pattern recognition in the programs could generate data on the genotype of the virus as well as the main contributor in samples with several viruses.

A testing for HPV-16 and HPV-18 is a common practice that was facilitated by the addition of the first two nucleotides as a program that showed AAC was identified as HPV-16 while ACC could identify HPV-18. Thus, the time of sequencing was reduced to a minimum. These primers could be designed according to the goal of a study to specifically identify a particular microbe that is prevalent in a region.

The major advantage of Pattern recognition in pyrosequencing is that it does not requiring the use of electrophoresis but instead bioluminescence (Gharizadeh *et al*, 2001).

Due to the use of general primers for amplification nonspecific binding to regions other than the target can lead to data that can become extremely puzzling to decipher. One solution is the use of nested PCR that would lengthen the protocol as well as not offer a solution with guarantee on the results. The use of primers that bind only the target region or species under study called target specific sequencing primers can offer an advantage. These primers are complementary to the sample under study to avoid the amplification of non-specific targets. Another advantage of this method is the circumventing of additional steps such as nested PCR or cloning operations (Gharizadeh *et al*, 2003b).

Thus, the use of such primers based on multiple-sequencing can circumvent the problem of such non-specific amplification.

ADDRESSING LOW STRENGTH SIGNALS

As compared to the overall product amplified, the sequence under study can occupy only a fraction of the total signal that causes its signal to be overlooked. Additionally, non-specific amplified products can also overshadow the DNA under study due to its lower level of amplification.

This can be resolved by using both multiple sequencing primers that target only specific regions under study and pyrosequencing that can read samples that have lower levels of DNA (Gharizadeh *et al*, 2003).

DESIGN OF PRIMERS

The presence of primer–primer annealing or the formation of hairpin loops cause peaks in the signals and hence require careful study to avoid such artefacts. These challenges can be overcome by the use of appropriate commercial software such as Assay Design Software from Biotage. An additional recommendation is the design of sequencing primers prior to PCR primers to facilitate the labelling of the appropriate strand using biotin (Gharizadeh *et al*, 2006).

ADVANCES IN SAMPLE PREPARATION

Sample preparation initially entailed the use of Vacuum Prep handset tool from Biotage, Uppsala, Sweden. A group in 2006 worked on the existing model to develop its efficiency in the following 3 aspects:

- The avoiding of aspiration of liquids in an uncontrolled manner that leads to wastage of reagents or templates.
- The efficiency of aspiration was to be enhanced.
- The entry of residual matter was to be avoided to avoid dilution.

Analysis of the Vacuum Prep handset tool revealed the presence of a larger than required container for liquid with a partition for sequestering, along with a single and narrow outlet that are accountable for the lack of control in the aspiration of liquids. An additional issue is the presence of residual liquids in the chamber that could lead to dilution of the samples. The presence of plates made of metal had the issue of rusting.

This model was worked upon at the Stanford Genome Technology Center, to make it into a tool that lacked the partition, thinner and aspiration outlets at either end of the equipment. The tool also was developed to have a filter pad to absorb residual liquids that would flow into the microplate wells. Further instructions are present at http://bioel.stanford.edu/vacuum_prep.

Pyrosequencing (PSQ HS96A) was used to compare the efficiencies of both the handsets for sample preparation of LS amplicon. The data from programs revealed that the handset developed by modification reported a decrease in loss of DNA as well as enhanced signals (Gharizadeh *et al*, 2006).

IMPROVEMENTS IN ENZYMES

The applications of pyrosequencing in the field of sequencing of genes as well as genomes involve the use of reads of longer lengths. This requires the use of enzymes with high stability as well as synchrony. To facilitate the addition of nucleotides in a repeated fashion, the removal of nucleotides from the reaction is warranted. The enzyme that is used to remove these nucleotides should be capable of performing the hydrolysis of all the deoxynucleoside triphosphates at similar rates as well as ATP. An additional feature is the rate at which the enzyme degrades nucleotides should be lesser than that of the polymerase enzyme that extends the DNA chain.

Table 3.1: Km of enzymes used for sequencing

Enzyme	Km
Klenow Polymerase	0.18
Apyrase	120
ATP sulfurylase	0.56
Firefly luciferase	20

The study of several systems has shown that apyrase optimally fulfils the requirements. This enzyme hydrolyzes the nucleotides at a value below the Km of the polymerase that results in incomplete polymerization. This warrants the use of repeated application of nucleotides or the incorporation of lesser quantities of apyrase in the system.

In the three-enzyme system used in solid-phase pyrosequencing, washing steps are critical that also yields an added benefit that includes a lack of substances that can inhibit the reaction due to the washing steps.

In the four-enzyme system, it has been reported that a decrease in dilution effect as well as abolishing the presence of inhibiting substances can yield a read of 200 nucleotides (Ronaghi *et al.*, 2000). Despite the high values of signal-to-noise ratios in the system, mis incorporation of nucleotides leads to a termination and hence a reduction in signal. This could involve either a minus frame shift where certain primers receive nucleotides behind that of others or a plus frame shift where few primers are ahead.

The use of the minus frame shift can be minimized by the use of exo DNA polymerase while it has been reported that plus frame shifts are the outcomes of improper degradation of nucleotides as well as the presence of contamination. This contamination can include nucleoside diphosphate kinase, present in commercial solutions of apyrase or sulfurylase (Karamohamed *et al.*, 1999).

The effect of dilution is another major factor in the four-enzyme system where the addition of nucleotides occurs cyclically to the reaction system. This results in an increase in reaction volume and consequent dilution in long read operations. The dilution of enzymes results in a dip in output that can be circumvented by facilitating evaporation by elevating the temperature or decreasing the volume of the system (Reviewed by Mostafa Ronaghi, 2001).

INTRODUCTION OF A PRIMER WASH STEP

The use of SSB in the pyrosequencing protocol does not entail the use of a washing step for the primers. However, certain studies have reported false peaks due to primer dimmers or secondary structures despite the presence of SSB (Gharizadeh et al., 2003a, 2005). This warrants the use of a primer wash that faces the following challenges:

- A loss of the beads that possess the single stranded DNA: This can be overcome by the use of more beads in the system and more volumes of PCR products.
- The probability of liquid left after washing that could lead to dilution can circumvented by the use of appropriate sample handling machines that prevent residual liquids.

The wash of primers has been recommended for multiplex sequencing primers as well as lack of flexibility when designing primers. This extra wash has been suggested to avoid excess as well as primers that have not bound the system though SSB is present in the system (Gharizadeh *et al*, 2006).

ADVANCES IN INSTRUMENTATION

Automated formats on the principle of micro-titre plates

This system employs the use of an inkjet cartridge that is disposable for sample application. The system can administer six different reagents on a micro-titer plate in a precise manner in volumes of 200 nL. This microtiter plate is agitated continuously to maintain constant parameters while the temperature of the system is kept constant. The requisite nucleotides and reagents are applied on the inket cartridge. Following the placement of the microtiter plate bearing the primer DNA, the nucleotides are added from the cartridge in a predetermined format.

Based on the principle of pyrosequencing, the luminescent signal generated from each well of the plate is collected and focused by a system of lenses to transfer the signal to a CCD camera. In order to lessen the signals emitted from each well from getting mixed up, the samples are administered in an alternating format. The reaction is monitored by the use of the CCD system that record images continuously. Several software systems are applied that include Windows NT based soft wares (Mostafa Ronaghi, 2001).

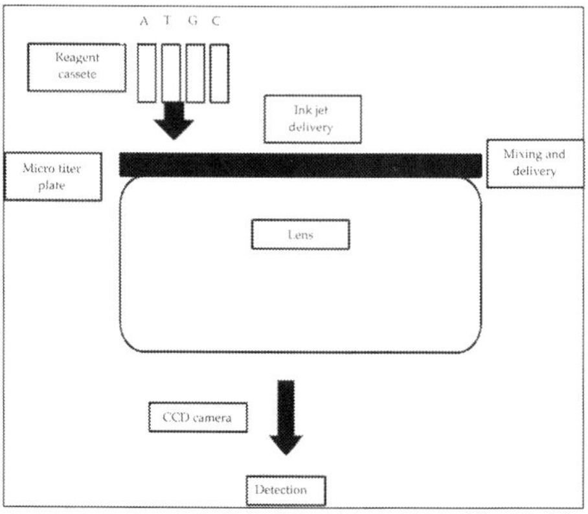

Figure 3.2: Automated formats on the principle of micro-titre plates.

Microfluidics Using Solid-Phase Pyrosequencing

The system involves immobilization of the templates on a solid support. The use of microfluidic systems in conjunction with a solid phase system can present advantages that include:

- Due to a decrease in cycling time, the speed of the sequencing can be enhanced.
- The process cost can be decreased by the use of lesser amounts of enzymes applied.
- A combination of amplification along with the sequencing can be applied in a single flow system.
- Due to wash steps following every cycle, there are no substances that inhibit the reaction

(Eckersten *et al*, 2000 and Mostafa Ronaghi, 2001).

Array Pyrosequencing

The applications of microarrays can be extended into the realm of sequencing as the combination of both technologies yields advantages that include: decreased costs as well as throughput. The use of both random as well as ordered arrays is possible in pyrosequencing of which the former circumvents the use of a step for decoding to yield a large amount of data pertaining to sequences. The efficient application of samples in a homogenous format can use a piezoelectric ultrasonic Sprayer. The use of fluorescence based technologies can employ the use of lasers for amplifying the signal as against chemiluminescent signals (Ronaghi, 2001).

Example of automated systems

454

An automated system based on liquid phase sequencing has extended the application of this technique for high throughput analysis (PSQ 96 systemhttp://www.pyrosequencing. com). 454 Life Sciences Corporation (Branford, CT, USA) uses a commercial application of microfluidics for pyrosequencing along with emulsion PCR systems. This system has reported an accuracy of 99% and can generate sequence reads of 100 bases and the sample number is 300,000. The procedure involves the shearing of genomic DNA followed by ligation with adapters. Using the principle of limiting dilution, one bead binds one fragment of DNA. A mixture of oil and PCR mixtures are then applied to the beads that is followed by amplification of the samples following breaking of the emulsions.

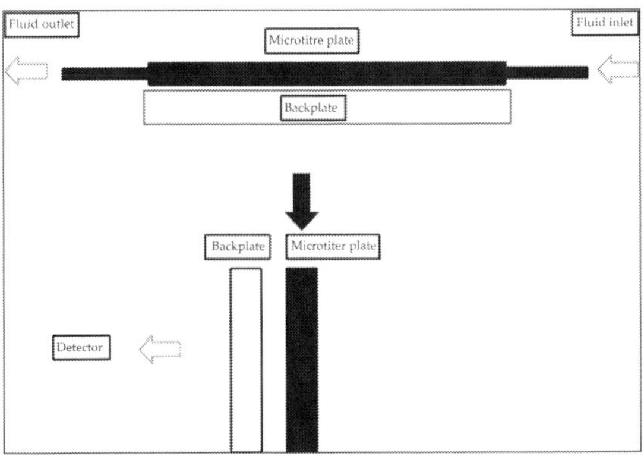

Figure 3.3: A schematic of 454 pyrosequencing.

The subsequent step is pyrosequencing that involves pyrosequencing on picotitre wells. The wells that have been framed on a fibre-optical slide can accommodate one bead. Follwing cyclic addition of reaction reagents the signals are captured by CCD (Fakruddin *et al*, 2012).

Machine for SNP used for pyrosequencing

A system that was applicable for analysis of SNPs: PSQ™HS96A DNA sequencing system was tested as a platform for pyrosequencing. The instrument has added advantages of Capillary Dispensation Tips that facilitate the utilization of lower volumes of products (a ten-fold reduction as compared to traditional instruments) as well as a CCD camera with enhanced sensitivity. The volume of samples required were decreased from 50 to 5 μl which can be reflected as a reduction in cost and effort involved in PCR reactions.

Quantitative studies showed that samples with low concentration of DNA can be sequenced by this modification. The limit of resolution of this modification was listed as 0.15 pmol as determined by NanoDrop spectrophotometer and 0.1 pmol as determined by exACTGene technique (Gharizadeh *et al*, 2006)

CASE STUDY: IMPROVEMENTS IN AND APPLICATION OF 454 PYROSEQUENCING FOR RARE HIV- 1 VARIANTS

454 system is reported as a massively parallel sequencing (MPS) approach employed in the study of genetic variation of HIV. A main challenge with pyrosequencing described is the appearance of errors by the creation of artifactual insertions or deletions (InDels) attributed to homopolymer runs. In order to meet this challenge, several algorithms have been employed that include: QuRe (Prosperi and Salemi, 2012) and ShoRAH (Zagordi *et al*, 2011). The mismatch rate in 454 has been listed as 0.02% that is lower than that when used in conjunction with an initial PCR amplification (Gilles et al, 2011).

A study in 2013 devised a strategy for the analysis of low frequency variants of HIV-1: using several approaches in sample preparation.

As the rate of mismatch in incorporation is dependent on the enzyme used: the team studied the effects of several enzymes:

- Advantage 2 (Adv2; Clontech, Mountain View, CA),
- Phusion High Fidelity DNA polymerase (New England Biolabs, Ipswich, MA),
- KOD Hot Start DNA polymerase (EMD Millipore, Billerica, MA) and
- Kapa HiFi Hot Start (Kapa Biosystems; Boston, MA).

It was reported that the lowest error rate was obtained when an enzyme with high fidelity was used in both rounds of amplification. The use of amplicons presented two advantages: firstly, the system can use the application of random bar codes on templates that avoids the errors of PCR or pyrosequencing. Secondly, the technique can utilize universal amplicons over the use of separate adaptors for the selection of DNA fragments. The authors designed 22 primers for the regions of *gag-pol* and *env* gp120 of HIV-1. The team reported that the detection of low level variants as 0.02-0.1%.: thus, increased detection of such variants (Larsen *et al*, 2013).

The 454-pyrosequencing technology has seen development with the years along with an enhancement in the reads it can perform. The sequence reads of the first-generation instrument (GS 20) is 60 Mb for an individual run and 100 bp. The sequence reads of the second-generation instrument (GS FLX) is to 150 Mb for an individual run while it is close to 500 Mb for an individual run for the third-generation (XLR and now GS FLX Titanium). The latest instrumentation: GS FLX+ has a read length of up to 1 kb (Siqueira *et al*, 2012).

CONCLUSION

The technique of pyrosequencing has been called as a user-friendly application that can be performed without complications. The technique does face certain challenges such as unwanted signal peaks or sequence reads or the challenge in reading homopolymer regions. However, appropriate measures have been reported through the use of additives or modifications of the technique to achieve optimal results. The technique has applications in several fields and hence, developments in various aspects of its use can aid in augmenting the application of this technology.

REFERENCES

1. Agaton C, et al., Gene expression analysis by signature pyrosequencing. Gene. 2002; 289:31– 9.
2. Ahmadian A, Ehn M, Hober S. Pyrosequencing: History, biochemistry and future. Clinica Chimica Acta. 2006; 363: 83 – 94.
3. Baback Gharizadeh. Method development and applications of Pyrosequencing technology. Doctoral dissertation from the Department of Biotechnology. 2003. Royal Institute of Technology, Stockholm, Sweden.
4. Eckersten A, O ¨ rlefors AE, Ellstro¨m C, Erickson A, Lo¨fman E, Eriksson A, et al., High-throughput SNP scoring in a disposable microfabricated CD device. In Proceedings of the Micro Total analysis Systems. 2000; Kluwer Academic Publishers.
5. Fakruddin MD, Chowdhury A, Nur Hossain MD, bin Mannan KS, Mazumdar RM. Pyrosequencing- principles and applications. Int J Life Science and Pharma Res. 2012; 2: 2.
6. Gharizadeh B, Akhras M, Nourizad N, et al., Methodological improvements of pyrosequencing technology. Journal of biotechnology. 2006; 124(3): 504-511. doi:10.1016/j.jbiotec. 2006.01.025.
7. Gharizadeh B, Kalantari M, Garcia CA, Johansson B, Nyren P. Typing of human papillomavirus by pyrosequencing. Lab Invest. 2001; 81:673-679.
8. Gharizadeh B, Nordstrom T, Ahmadian A, Ronaghi M, Nyren P. Long-read pyrosequencing using pure 2'-deoxyadenosine5'-O'-(1-thiotriphosphate) Sp-isomer. Anal Biochem. 2002; 301: 82-90.
9. Gharizadeh, B., Ghaderi, M., Donnelly, D., Amini, B., Wallin, K.L. and Nyren, P.: Multiple-primer DNA sequencing method. Electrophoresis. 2003b; 24: 1145-1151.
10. Gilles A, Meglécz E, Pech N, Ferreira S, Malausa T et al., Accuracy and quality assessment of 454 GS-FLX Titanium pyrosequencing. BMC Genomics. 2011; 12: 245. doi:10.1186/1471-2164-12-245.
11. Karamohamed S, Nilsson J, Nourizad K, Ronaghi M, Pettersson B, Nyren P. Production, purification, and luminometric analysis of recombinant Saccharomyces cerevisiae MET3 adenosine

triphosphate sulfurylase expressed in Escherichia coli. Prot Exp Purif. 1999; 15: 381–388.

12. Larsen BB, Chen L, Maust BS, Kim M, Zhao H, et al., Improved Detection of Rare HIV-1 Variants using 454 Pyrosequencing. PLoS ONE. 2013; 8(10): e76502. doi:10.1371/journal.pone.0076502.
13. Meyer RR, Laine PS. The single-stranded DNA-binding protein of Escherichia coli. Microbiol Rev. 1990; 54:342–380.
14. Mostafa Ronaghi. Pyrosequencing Sheds Light on DNA Sequencing. Genome Res. 2001 11: 3-11.
15. Nordstrom T, Gharizadeh B, Pourmand N, Nyren P, Ronaghi M. Method enabling fast partial sequencing of cDNA clones. Anal Biochem. 2001; 292: 266-71.
16. Nourizad, N., Gharizadeh, B. and Nyren, P. Method for clone checking. Electrophoresis. 2003; 24:1712-1715.
17. Prosperi MC, Salemi M. QuRe: software for viral quasispecies reconstruction from next-generation sequencing data. Bioinformatics. 2012; 28: 132-133. doi:10.1093/bioinformatics/btr627.
18. Ronaghi M, Pourmand N, Jain M, Willis T, Davis R. Pyrosequencing for genome resequencing. In 12th International Genome Sequencing and Analysis Conference, 2000. Miami, FL.
19. Ronaghi, M., Karamohamed, S., Pettersson, B., Uhlen, M. and Nyren, P.: Real-time DNA sequencing using detection of pyrophosphate release. Anal Biochem. 1996; 242: 84-89.
20. Siqueira JF, Fouad AF, Rôças IN. Pyrosequencing as a tool for better understanding of human microbiomes. J Oral Microbiol. 2012; 4: 10.3402/jom.v4i0.10743. doi:10.3402/jom. v4i0.10743.
21. Zagordi O, Bhattacharya A, Eriksson N, Beerenwinkel N. ShoRAH: estimating the genetic diversity of a mixed sample from next generation sequencing data. BMC Bioinfo. 2011; 12: 119. doi: 10.1186/1471-2105-12-119. PubMed: 21521499.

CHAPTER 4

APPLICATIONS OF PYROSEQUENCING

Shiv Sanjeevi Sripathi[1] and Prerna Pandey[2]

[1]Shiv Sanjeevi Sripathi completed his BSc in Biotechnology from Mumbai University and secured second rank in Mumbai University and first rank in his college: Kishinchand chellaram College. Shiv secured first rank in his college KET's V.G.Vaze College and has worked on a stem cell project at the Specialized Centre for Cell Based Therapy (SCCT), KEM Hospital at Mumbai, following which he was a Junior Research Fellow at Microbiology & Cell Biology Department, Indian Institute of Science, Bangalore from August 2009 to November 2011. The field of work involved cloning of cell wall genes and transcription factors in E. coli & M. smegmatis. This was followed by a stint as a Research and Development Executive at Transasia Biomedicals, Mumbai where he worked on patent writing and Immunoassays from 2011 till 2013. In 2013, he entered the teaching field where he has been tutoring biology and chemistry to the students of 11th and 12th grades along with English grammar and communication skills to people of all age groups. Shiv also loves helping people write scientific manuscripts.

[2]Dr Prerna Pandey, a biotechnologist with several years of wet lab research experience, worked at International Center for Genetic Engineering and Biotechnology, New Delhi. Her field of research in PhD included isolation and molecular characterization of geminiviruses, genome sequencing, gene annotation, and gene silencing using the RNA interference technology. She has also worked at Transasia Biomedicals and Advance Enzyme Technologies as a scientist. Prerna has published papers in peer-reviewed journals, and has submitted a number of annotated Geminiviral genome sequences in the GenBank, including two novel ones. She has also completed her editing and proof-

reading courses from Society for the promotion of Editors and Proof-readers, UK and now works as a freelance scientific editor and writer. When Prerna is not busy with her assignments, she loves to explore historical places.

INTRODUCTION

Today DNA sequencing is one of the most important platforms for the study of biological systems. Genetics research has benefited enormously from the release of the human genome sequence. Subsequent technology has been developed and adapted to accommodate the need for faster, easier throughput of genetic assays. Pyrosequencing is a unique system that allows the analysis of genetic variations including single-nucleotide polymorphisms, indels and short repeats, as well as assessing RNA allelic imbalance, DNA methylation status, and gene copy number. Advances in methodology, including multiplex and universal primer applications, that have reduced cost for assay and also improved the throughput.

Pyrosequencing technology is a novel DNA sequencing technology, which had been developed at the Royal Institute of Technology (KTH), and is the first alternative to the conventional Sanger method for de novo DNA sequencing.

This method relies on the luminometric detection of pyrophosphate that is released during primer-directed

DNA polymerase catalyzed nucleotide incorporation. It is suited for DNA sequencing of up to one hundred bases and it offers a number of unique advantages (Gharizadeh et al., 2003a). This technique is a widely applicable, alternative approach for the detailed characterization of nucleic acids. Pyrosequencing has potential advantages of accuracy, flexibility parallel processing, and can be easily automated. Furthermore, the technique avoids the need for labeled primers, labeled nucleotides, and gel electrophoresis. Pyrosequencing has been successful for both confirmatory sequencing and denovo sequencing (Ronaghi, 2001).

Pyrosequencing technology also provides a large number of sequence reads in a single run, resulting in very large sampling depth and allowing detection not only of the most dominant microbial community members but also of the low-abundance (rare) taxa (Sogin et al.,2006). Pyrosequencing of the 16S rRNA gene from diverse environments has demonstrated that the microbial diversity can be orders of magnitude higher than appreciated by previous technologies (Kunin et al.,2010).

Principles and automation of Pyrosequencing

The availability of an automated system for liquid-phase pyrosequencing (PSQ 96 system, http://www.pyrosequencing. com) has allowed the technique to be adapted for high-throughput analyses (Ronaghi, 2001). Pyrosequencing is now being applied in microfluidic format commercially by 454 Life Sciences Corporation (Branford, CT, USA). The microfluidic pyrosequencing has been integrated with emulsion PCR and DNA sequencing assembly software. Sequence reads of up to 100 bases are generated, allowing massive parallel fashion sequencing in picolitre volumes. The new platform has the capacity to sequence up to 300,000 samples and generate up to 20-40 million bases at an accuracy of 99% per 4 hours sequencing run. In brief, the entire genomic DNA is sheared, and the fragments are ligated to adapters. The fragments are bound to the beads that favour one fragment per bead by limiting dilution approach. The beads are captured in the droplets of a PCR-reaction-mixture-in oil (Margulies et al., 2005). The DNA fragments are clonally amplified and the emulsions are broken after complete amplification. After single-strand treatments, the beads are deposited into the picotitre wells (etched in a fibre-optic slide). Each well has the capacity for a single bead. Smaller beads carrying immobilized enzymes for pyrosequencing reaction are added to the wells and the reagents flow cyclically through the wells. The generated photons are detected by a CCD camera and presented in flowgrams by integrated software (Gharizadeh et al., 2007).

APPLICATIONS OF PYROSEQUENCING

Pyrosequencing has opened up new possibilities for performing sequence-based DNA analysis (Ronaghi, 2001). Pyrosequencing is well suited for de novo sequencing and resequencing (Ronaghi, 2001). Currently, pyrosequencing method is broadly being used in many applications such as Single Nucleotide Polymorphism (SNP) genotyping (Ahmadian et al., 2000a; Nordstrom et al., 2000; Milan et al., 2000), identification of bacteria (Gharizadeh et al., 2003a; Grahn et al., 2003; Jonasson et al., 2002), fungal (Gharizadeh et al.,2005; Trama et al., 2005a; Trama et al.,

2005b) and viral typing (Gharizadeh et al., 2001; Adelson et al.,2005; Elahi et al., 2003; Gharizadeh et al., 2003a; Gharizadeh et al., 2005). Moreover, the method has demonstrated the ability to determine difficult secondary structures (Ronaghi et al., 1999) and perform mutation detection (Ahmadian et al.,2000b; Garrcia et al., 2000), DNA methylation analysis (Neve et al., 2002; Rickert et al., 2002; Uhlmann et al., 2002), multiplex sequencing (Gharizadeh et al., 2003b; Gharizadeh et al., 2006), tag sequencing of cDNA library (Nordstrom et al., 2001) and clone checking (Nourizad et al., 2003). Another highly significant application is whole genome sequencing (Margulies et al., 2005).

Genotyping of Single-Nucleotide Polymorphisms

For analysis of single-nucleotide polymorphisms (SNPs) by pyrosequencing, the 3'-end of a primer is designed to hybridize one or a few bases before the polymorphic position. In a single tube, all the different variations can be determined as the region is sequenced. A striking feature of program readouts for SNP analysis is the clear distinction between the various genotypes; each allele combination (homozygous or heterozygous) will give a specific pattern compared to the two other variants (Ahmadian et al., 2000a; Alderborn et al.,2000; Ekstrom et al., 2000; Nordstrom et al., 2000). This feature makes typing extremely accurate and easy. Relative standard deviation values for the ratio between key peaks of the respective SNPs and reference counterparts are $\leq$0.1 (Alderborn et al.,2000). Simple manual comparison of predicted SNP patterns and the raw data obtained can score an SNP, especially as no editing is needed. Because specific patterns can be readily achieved for individual SNPs, it is also be possible to automatically score the allelic status by pattern recognition software (Ronaghi, 2001).

As pyrosequencing signals are very quantitative, it is possible to use this strategy for the studies of allelic frequency in large population. This system allows >5000 samples to be analyzed in 8 h. Furthermore, pyrosequencing enables determination of the phase of SNPs when they are in the vicinity of each other allowing the detection of haplotypes (Ahmadian et al., 2000b).

Resequencing

Pyrosequencing is currently the fastest method for sequencing a PCR product. Because pyrosequencing generates an accurate quantification of the mutated nucleotides, the resequencing of PCR-amplified disease associated genes for mutation scanning will be one of the interesting applications. Using this technique for resequencing results in longer read length than de novo sequencing because nucleotide delivery can be specified according to the order of the sequence (Ronaghi, 2001). Programmed dispensing generates a signal for each addition in a pyrogram, therefore variation in the pattern indicates the appearance of a mutation. This strategy has been used for resequencing of the p53 tumor suppressor gene, where mutations were successfully determined and quantified (Garrcia et al., 2000).

Tag Sequencing

The sequence order of nucleotides determines the nature of the DNA. Theoretically, eight or nine nucleotides in a row should define a unique sequence for every gene in the human genome. However, it has been found that, to uniquely identify a gene of a complex organism such as human, a longer sequence of DNA is needed. In a pilot study, 98% of genes in a human cDNA library could be uniquely identified by sequencing a length of 30 nucleotides. Pyrosequencing was used to sequence this length for gene identification and the results were in complete agreement with longer sequence data obtained by Sanger DNA sequencing.

Pyrosequencing offers high-throughput analysis of cDNA libraries because 96 samples can be analyzed in less than one hour. Like Sanger DNA sequencing, pyrosequencing also has the advantage of library screening as the original cDNA clone is directly available for further analysis (Ronaghi, 2001).

Analysis of Difficult Secondary Structures

Hairpin structures are common features in genomic materials and have been proposed to have regulatory functions in gene transcription and replication. However, analyzing these sequences by conventional DNA sequencing usually gives rise to DNA sequence ambiguities seen as "run-off" or compressions. These problems have been associated with gel electrophoresis (Ronaghi, 2001). Pyrosequencing was successfully applied to decipher the sequence of such regions (Ronaghi et al., 1999).

Microbial and viral typing

Near instantaneous detection of pathogens from clinical materials is important for diagnosis, treatment and prophylactic measurements. The Microbial and viral threats have been altered in the course of history as a substantial number of pathogens have been eradicated or controlled. New pathogens are emerging and some are developing resistance. Accurate and specific typing of microbial and viral pathogens is of utmost importance in clinical diagnosis (Gharizadeh et al., 2003a). Many microorganisms usually lack adequate morphological details for easy identification. Furthermore, the development of appropriate therapies/ vaccines requires implementation of sufficient parameters for microbial and viral detection, and a reliable and robust genotyping method is necessary for accurate follow-up during clinical trials and serveillance of treatment (Van Doorn et al., 2001). Pyrosequencing technique, due to its many advantages, has been turned out to be an useful implement for microbial and viral typing. It is currently being applied for rapid typing of a large number of bacteria, yeasts, and viruses (Ronaghi, 2001).

Fungal identification

Pyrosequencing appears to be a good diagnostic tool for detection and identification of fungal pathogens. It can be utilized in typing of fungi clinically isolated from immune-compromised patients suffering from proven invasive fungal infections. For typing, the DNA has to be amplified by general consensus primers (Einsele et al., 1997) having complementary to a highly conserved region within the 18S rRNA gene allowing the amplification of a broad range of fungal species. The amplified DNA fragments are then sequenced up to 40 bases to identify the samples. All the samples can be accurately identified by the pyrosequencing technique. The reproducibility of the technique has been confirmed by sequencing the amplicons several times with identical sequence data each time for every sample. The sequence data obtained by pyrosequencing technique suggested that 18 to 32 bases are sufficient for identification of *C. albicans, C. glabrata, C. krusei, C. parapsilosis, C. tropicalis and Aspergillus* spp. (Gharizadeh et al., 2003b).

Bacterial identification

DNA markers used for typing normally contain both conserved and variable regions. A DNA primer complementary to the conserved or semiconserved region is usually employed to sequence the variable region. In bacteria, 16s rRNA gene is commonly used to identify different species and strains. By analyzing a sequence between 20–100 nucleotides on 16S rRNA gene, it is possible to group different bacteria taxonomically and, in many cases to get information about strains. Further development of bacterial identification by pyrosequencing could be brought about in order for the method to be suitable for regions other than 16s rRNA of different bacteria (Gharizadeh et al., 2003a).

Fungal identification

Pyrosequencing appears to be a good diagnostic tool for detection and identification of fungal pathogens. It can be utilized in typing of fungi clinically isolated from immune-compromised patients suffering from proven invasive fungal infections. For typing, the DNA has to be amplified by general consensus primers (Einsele et al., 1997) having complementary to a highly conserved region within the 18S rRNA gene allowing the amplification of a broad range of fungal species. The amplified DNA fragments are then sequenced up to 40 bases to identify the samples. All the samples can be accurately identified by the pyrosequencing technique. The reproducibility of the technique has been confirmed by sequencing the amplicons several times with identical sequence data each time for every sample. The sequence data obtained by pyrosequencing technique suggested that 18 to 32 bases are sufficient for identification of C. albicans, C. glabrata, C. krusei, C. parapsilosis, C. tropicalis and Aspergillus spp. (Gharizadeh et al., 2003b).

Epigenetics

Hypermethylation of CpG islands located within or close to the 5' region of genes is associated with inhibition of gene expression. This can be particularly important in tumors where hypermethylation can "switch off" tumor-suppressor genes and/or DNA damage repair genes, which can lead to sensitivity or resistance to chemotherapy drugs (Teodoridis et al.,2005). Assessment of methylation status in CpG islands can be easily performed using bisulfate modification and comparing the conversion of cytidine to thymidine with the allele quantification Pyrosequencing software. The procedure for detecting methylation using Pyrosequencing (Tost et al.,2003) was recently applied to determine the methylation status in oral squamous cell carcinoma (Shaw et al.,2006), leading to the identification of cytoglobin as a novel putative tumor suppressor gene.

Medical Genetics

The use of Pyrosequencing to rapidly and accurately type HLA loci is of valuable clinical importance (Entz et al.,2005; Wu et al.,2004; Ringquist et al.,2004; Ramon et al.,2003; Hochberg et al.,2003; Ringquist et al.,2002). The matching of patient and donor for HLA significantly reduces the chance of rejection following organ transplantation.

Other medical applications for Pyrosequencing include the typing of mutations conferring antibiotic resistance to bacteria or to differentiate between multiple bacterial or viral strains (Innings et al.,2005; Pai et al.,2005; Sinclair et al.,2005) and for screening for target phage ligands for novel drug development (Rahim et al.,2003). These protocols demonstrate the utility of Pyrosequencing for clinical assessment and drug development applications.

Pyrosequencing to analyze HIV diversity

The pyrosequencing methods are well suited to addressing questions on the dynamics of HIV quasispecies in response to selective pressures. HIV reverse transcriptase is very error prone, making roughly one base pair substitution mutation per round of replication (Preston,1997). The viral populations in infected individuals are also very large, with some 1010 virions produced and destroyed per day (Coffin,1995; Prelson et al.,1996). After infection, this activity quickly results in the formation of a diverse pool, or quasispecies, where most viral sequences differ from all others.

Pyrosequencing offers an improved means of characterizing sequence variation present in such large populations. One application is quantification of rare drug resistance mutations in treated individuals failing anti-retroviral therapy. Current guidelines recommend that whenever an HIV-positive individual is initiating or changing therapy, possible drug resistance alleles should be assayed, and treatment choices adjusted accordingly. However, the most commonly used genotyping methods only provide information on the most abundant sequence variants. Evidence suggests that rare drug resistance variants, when subjected to the selective pressures of drug treatment, can quickly grow out and become the predominant form, leading to treatment failure (Palmer et al.,2005; Jourdain et al.,2004). Two reports have applied pyrosequencing to the detection and characterization of rare drug resistant variants (Hoffmann et al.,2007). Shafer and colleagues (Wang et al.,2007) purified RNA from patient virions, reverse transcribed to generate cDNA, sheared the cDNA product, and carried out pyrosequencing as described for mammoth DNA. Hoffmann et al., took a different approach, PCR amplifying short regions of interest, then pyrosequencing the amplicons directly. Both groups also analyzed control cloned viral stocks, allowing empirical determination of the rates of false positive calls and thereby permitting convincing demonstrations of detection of drug resistance mutations present as <1% of the population.

Another study applied pyrosequencing to analyzing viral tropism in quasispecies inferred from V3 loop sequences (Tsibris et al.,2006). As inhibitors that block coreceptor binding come into widespread use, there will be increased need for deep profiling of coreceptor usage prior to initiating therapy, and pyrosequencing would be an ideal tool for this application.

Pyrosequencing to characterize unknown opportunistic infections

Clinicians are occasionally confronted with apparent infections that cannot be easily attributed to a known pathogen. Given concerns about bioterrorism, there is an urgent need to identify the infectious agents quickly in such cases. For AIDS patients, opportunistic infections that would be cleared in immunocompetent individuals can cause morbidity in the immunocompromised, potentially resulting in challenges in identifying the agent. A wealth of new molecular methods are becoming available for identifying new pathogens, with pyrosequencing a major addition.

The discovery of a virus potentially responsible for colony collapse disorder (CCD) of honey bees provides a striking example (Cox et al.,2007). North American bee colonies have been failing at alarming rates, and an infectious agent implicated. Pyrosequencing of large numbers of samples from affected and unaffected colonies revealed that the presence of Israeli acute paralysis virus was strongly associated with CCD, and follow-up quantitative PCR studies strengthened the connection. Experimental infection studies are now needed to test the causalties. Pyrosequencing also provides potent new methods for analyzing bacterial populations. Many bacterial diseases likely involve not only single pathogens but the full microbial community in which that pathogen resides--for example, obesity and Crohn's disease are proposed to involve community-wide alterations in GI microbiota (Manichanh et al.,2006; Turnbaugh et al.,2006). Pyrosequencing of DNA samples from uncultured bacterial communities can identify the members of a community and their relative abundance in a rapid and cost effective fashion (Sogin et al.,2006). For example, 454 pyrosequencing of segments of the 16S RNA genes from uncultured communities of marine bacteria has revealed numerous previously undiscovered bacterial taxa (Huber et al.,2007).

Pyrosequencing in hypothesis testing

Pyrosequencing can be used not just to sequence genomes, but also as an end-point assay in a mechanistic experiment, as one might once have used a p24 assay or Southern blot. For example, in an application in the integration targeting field, HIV integration in cells containing or lacking the targeting factor PSIP1/LEDGF/p75 was compared using pyrosequencing. The analysis revealed that cells lacking the cofactor showed reduced integration in transcription units.

In another example, Shuman and colleagues identified mitoxantrone as an anti-poxviral agent, and sought to identify the viral target of the inhibitor (Deng et al.,2007). They picked mutant vaccinia viruses insensitive to the drug, then they sequenced the entire 195 kb mutant genomes from mutants and compared them to the wild type. A mutation in the ligase gene was found to be selectively present in the genomes of the drug resistant variants. Thus pyrosequencing allowed efficient identification of the molecular target of a new antiviral agent in a large viral genome.

PROSPECTS OF PYROSEQUENCING

Future applications require more robust and efficient DNA sequencing techniques for sequence determination. The Pyrosequencing method has already shown evidence of high accuracy in DNA sequencing and analysis of polymorphic DNA fragments in many clinical and research settings. It is a relatively straightforward and user-friendly method possessing unique methodological characteristics and this technique is currently being used in multidisciplinary fields in academic, clinical and industrial settings. By increasing the read length to higher scores and by shortening the sequence reaction time per base calling, pyrosequencing may take over many broad areas of DNA sequencing applications as the trend is directed to analysis of fewer amounts of specimens and large-scale settings, with higher throughput and lower cost (Gharizadeh *et al.*, 2007).

Pyrosequencing has shown excellent accuracy in analysis of polymorphic DNA fragments. This technology has also been used for quantification of allelic frequency in populations. Pyrosequencing will have a large impact in that area because a large number of samples can be pooled in one pyrosequencing reaction. A high throughput version of this technology can potentially be used for resequencing o genomes. Pyrosequencing technology is relatively new and there lies ample room for versatile developments in both chemistry and instrumentation. This technology is already time-and cost competitive as compared to the most conventional sequencing methods. Work is underway to further improve the chemistry, to measure the sequencing efficiency at elevated temperatures and to run the reaction in miniaturized formats (Ronaghi, 2001).

SUMMARY

The availability of a multipurpose instrument for genetic analysis is a major boon to researchers. Pyrosequencing is an established, reliable technique that can be applied to all areas of genetics. Applications of

Pyrosequencing technology can be found in widely diverse research fields from plant biology to immunology. There are still opportunities for further breakthroughs in this area as only very few research groups have focused on this method. In future, pyrosequencing technology is expected to achieve longer read length, to reduce the sequencing time frame, to decrease the sample quantity and to make further improvements in automation.

REFERENCES

1. Adelson ME, Feola M, Trama J, Tilton RC, Mordechai E. Simultaneous detection of herpes simplex virus types 1 and 2 by realtime PCR and pyrosequencing. J. Clin. Virol. 2005; 33: 25-34.
2. Ahmadian A, Gharizadeh B, Gustafsson AC, Sterky F, Nyren P, Uhlen M, Lundeberg J. Single nucleotide polymorphism analysis by pyrosequencing. Anal. Biochem. 2000a; 280: 103–110.
3. Ahmadian A, Lundeberg J, Nyren P, Uhlen M, Ronaghi M. Analysis of the p53 tumor suppressor gene by pyrosequencing. Biotechniques. 2000b; 28: 140–147.
4. Alderborn A, Kristofferson A, Hammerling U. Determination of single nucleotide polymorphisms by real-time pyrophosphate DNA sequencing. Genome Res. 2000; 10: 1249–1258.
5. Coffin JM. HIV population dynamics in vivo: implications for genetic variation, pathogenesis, and therapy. Science. 1995; 267:483–486.
6. Cox-Foster DL, Conlan S, Holmes EC, Palacios G, Evans JD, Moran NA, et al. A metagenomic survey of microbes in honey bee colony collapse disorder. Science. 2007; 318:283–287.
7. Deng L, Dai P, Ciro A, Smee DF, Djaballah H, Shuman S. Identification of novel antipoxviral agents: mitoxantrone inhibits vaccinia virus replication by blocking virion assembly. J Virol. 2007; 81:13392–13402.
8. Einsele H, Hebart H, Roller G, Loffler J, Rothenhofer I, et al. Detection and identification of fungal pathogens in blood by using molecular probes. J. Clin. Microbiol.1997; 35: 1353-1360.
9. Ekstrom B., Alderborn A. and Hammerling U. Pyrosequencing for

SNPs. Progress Biomed. Optics. 2000; 1: 134–139.

10. Elahi E, Pourmand N, Chaung R, Rofoogaran A, Boisver J, et al. Determination of hepatitis C virus genotype by pyrosequencing. J. Virol. Methods. 2003; 109: 171-176.
11. Entz P, Toliat MR, Hampe J, et al. New strategies for efficient typing of HLA class-II loci DQB1 and DRB1 by using Pyrosequencing. Tissue Antigens. 2005; 65: 67–80.
12. Espinosa V, Kettlun AM, Zanocco A, Cardemil E, Valenzuela MA. Differences in nucleotide-binding site of isoapyrases deduced from tryptophan fluorescence. Photochemistry. 2003; 63:714.
13. Garrcia CA, Ahmadian A, Gharizadeh B, Lundeberg J, Ronaghi M, Nyren P. Mutation detection by pyrosequencing:Sequencing of exons 5-8 of the p53 tumour suppressor gene. Gene. 2000; 253: 249-257.
14. Gharizadeh B, Kalantari M, Garcia CA, Johansson B, Nyren P. Typing of human papillomavirus by pyrosequencing. Lab. Invest. 2001; 81: 673-679.
15. Gharizadeh B, Kaller M, Nyren P, Andersson A, Uhlen M, Lundeberg J, Ahmadian A. Viral and microbial genotyping by a combination of multiplex competitive hybridization and specific extension followed by hybridization to generic tag arrays. Nucl. Acids Res. 2003a; 31: e146.
16. Gharizadeh B, Ghaderi M, Donnelly D, Amini B, Wallin KL, Nyren P. Multiple primer DNA sequencing method. Electrophoresis. 2003b; 24: 1145-1151.
17. Gharizadeh B, Oggionni M, Zheng B, Akom E, Pourmand N, et al. Type-specific multiple sequencing primers: A novel strategy for reliable and rapid genotyping of human papillomaviruses by pyrosequencing technology. J. Mol. Diagn. 2005; 7: 198-205.
18. Gharizadeh B, Zheng B, Akhras M, Ghaderi M, Jejelowo O, et al., Sentinel-base DNA genotyping using multiple sequencing primers for high-risk human papillomaviruses. Mol. Cell Probes. 2006; 20: 230-238.
19. Gharizadeh B, Ghaderi M, Nyren P. Pyrosequencing technology for short DNA sequencing and whole genome sequencing. Tech. 2007; 47: 129-132.

20. Grahn N, Olofsson M, Ellnebo-Svedlund K, Monstein HJ, Jonasson J. Identification of mixed bacterial DNA contamination in broad range PCR amplification of 16S rDNA V1 and V3 variable regions by pyrosequencing of cloned amplicons. FEMS Microbiol. Lett. 2003; 219: 87–91.

21. Hochberg, E. P., Miklos, D. B., Neuberg, D., et al. A novel rapid single nucleotide polymorphism (SNP) based method for assessment of hematopoietic chimerism after allogeneic stem cell transplantation. Blood. 101; 2003: 363–369.

22. Hoffmann C, Minkah N, Leipzig J, Wang G, Arens MQ, Tebas P, et al. DNA bar coding and pyrosequencing to identify rare HIV drug resistance mutations. Nucleic Acids Res. 2007; 35:e91.

23. Huber JA, Welch DB, Morrison HG, Huse SM, Neal PR, Butterfield DA, et al. Microbial population structures in the deep marine biosphere. Science. 2007; 318: 97–100.

24. Innings A, Krabbe M, Ullberg M, Herrmann B. Identification of 43 Streptococcus species by pyrosequencing analysis of the rnpB gene. J. Clin. Microbiol. 2005; 43, 5983–5991.

25. Jonasson J, Olofsson M, Monstein HJ. Classification, identification and subtyping of bacteria based on pyrosequencing and signature matching of 16S rDNA fragments. APMIS. 2002; 110: 263–272.

26. Jourdain G, Ngo-Giang-Huong N, Le Coeur S, Bowonwatanuwong C, Kantipong P, Leechanachai P, et al. Intrapartum exposure to nevirapine and subsequent maternal responses to nevirapine based antiretroviral therapy. N. Engl. J. Med. 2004; 351: 229–240.

27. Karamohamed S, Nilsson J, Nourizad K, Ronaghi M, Pettersson B, et al. Production, purification and luminometric analysis of recombinant Saccharomyces cerevisiae MET3 adenosine triphosphate sulfurylase expressed in Escherichia coli. Protein Exp. Purif. 1999; 15: 381-388.

28. Kunin V, Engelbrektson A, Ochman H, Hugenholtz P. Wrinkles in the rare biosphere: pyrosequencing errors can lead to artificial inflation of diversity estimates. Environ. Microbiol. 2010; 12: 11823.

29. Manichanh C, Rigottier-Gois L, Bonnaud E, Gloux K, Pelletier E, Frangeul L, et al. Reduced diversity of faecal microbiota in Crohn's disease revealed by a metagenomic approach. Gut. 2006; 55:205–

211.

30. Margulies M, Egholm M, Altman WE, Attiya S, Bader JS, et al. Genome sequencing in microfabricated high-density picolitre reactors. Nature. 2005; 437: 376-380.
31. Milan D, Jeon JT, Looft C, Amarger V, Robic A, et al. A mutation in PRKAG3 associated with excess glycogen content in pig skeletal muscle. Science. 2000; 288: 1248-1251.
32. Neve B, Froguel P, Corset L., Vaillant E, Vatin V, Boutin P. Rapid SNP allele frequency determination in genomic DNA pools by pyrosequencing. Biotechniques. 2002; 32: 1138-1142.
33. Nordstrom T, Ronaghi M, Forsberg L, de Faire U, Morgenstern R, Nyren P. Direct analysis of single-nucleotide polymorphism on double-stranded DNA by pyrosequencing. Biotechnol. Applied Biochem. 2000; 31: 107-112.
34. Nordstrom T, Gharizadeh B, Pourmand N, Nyren P, Ronaghi M. Method enabling fast partial sequencing of cDNA clones. Anal. Biochem. 2001; 292: 266-271.
35. Nourizad N, Ehn M, Gharizadeh B, Hober S, Nyren P. Methylotrophic yeast Pichia pastoris as a host for production of ATP diphosphohydrolase (apyrase) from potato tubers (Solanum tuberosum). Protein Exp. Purif.,2003: 27: 229-237.
36. Nourizad N., Gharizadeh B. and Nyren P. Method for clone checking. Electrophoresis. 2003; 24: 1712-1715.
37. Nyren P, Lundin A. Enzymatic method for continuous monitoring of inorganic pyrophosphate synthesis. Anal. Biochem. 1985; 151: 504-509.
38. Nyren P. Apyrase immobilized on paramagnetic beads used to improve detection limits in bioluminometric ATP monitoring. J. Bioluminescence Chemiluminescence. 1994; 9: 29-34.
39. Nyren P, Skarpnack SE. Method of sequencing DNA based on the detection of the release of pyrophosphate and enzymatic nucleotide degradation. 2001: U.S. Patent 6258568.
40. Palmer S, Kearney M, Maldarelli F, Halvas EK, Bixby CJ, Bazmi H, et al. Multiple, linked human immunodeficiency virus type 1 drug resistance mutations in treatment-experienced patients are missed

by standard genotype analysis. J. Clin. Microbiol. 2005; 43:406–413.

41. Pai R, Limor J, Beall B. Use of pyrosequencing to differentiate Streptococcus pneumoniae serotypes 6A and 6B. J. Clin. Microbiol. 2005; 43, 4820–4822.
42. Perelson AS, Neumann AU, Markowitz M, Leonard JM, Ho DD. HIV-1 Dynamics in Vivo: Virion Clearance Rate, Infected Cell LifeSpan, and Viral Generation Time. Science. 1996; 271:1582–1586.
43. Preston BD. Reverse transcriptase fidelity and HIV-1 variation. Science. 1997; 275:228–229.
44. Rahim A, Coutelle C, Harbottle R. High-throughput Pyrosequencing of a phage display library for the identification of enriched target-specific peptides. Biotechniques. 2003; 35, 317–324.
45. Ramon D, Braden M, Adams S, Marincola FM, Wang L. Pyrosequencing trade mark: A one-step method for high resolution HLA typing. J. Transl. Med. 2003; 1, 9.
46. Rickert AM, Premstaller A, Gebhardt C, Oefner PJ. Genotyping of Snps in a polyploidy genome by pyrosequencing. Biotechniques. 2002; 32: 592-600.
47. Ringquist S, Alexander AM, Rudert WA, Styche A, Trucco M. Pyrosequence-based typing of alleles of the HLA-DQB1 gene. Biotechniques. 2002; 33, 166–165.
48. Ringquist S, Alexander AM, Styche A, Pecoraro C, Rudert WA, Trucco M. HLA class II DRB high resolution genotyping by pyrosequencing: comparison of group specific PCR and pyrosequencing primers. Hum. Immunol. 2004; 65, 163–174.
49. Ronaghi M, Karamohamed S, Pettersson B, Uhlen M, Nyren P. Real-time DNA sequencing using detection of pyrophosphate release. Anal. Biochem. 1996; 242: 84–89.
50. Ronaghi M, Uhlen M, Nyren P. A sequencing method based on real-time pyrophosphate. Science. 1998; 281: 363- 365.
51. Ronaghi M, Nygren M, Lundeberg J, Nyren P. Analyses of secondary structures in DNA by pyrosequencing. Anal. Biochem. 1999; 267: 65-71.

52. Ronaghi M. Pyrosequencing sheds light on DNA sequencing. Genome Res. 2001; 11: 3-11.

53. Shaw RJ, Liloglou T, Rogers SN, et al. Promoter methylation of P16, RARbeta, E-cadherin, cyclin A1 and cytoglobin in oral cancer: quantitative evaluation using pyrosequencing. Br. J. Cancer. 2006; 94, 561–568.

54. Sinclair, A., Arnold, C., and Woodford, N. Rapid detection and estimation by pyrosequencing of 23S rRNA genes with a single nucleotide polymorphism conferring linezolid resistance in Enterococci. Antimicrob. Agents Chemother.2003: 47,3620–3622.

55. Siqueira JF, Jr, Rocas IN. Molecular analysis of endodontic infections. In: Fouad AF, ed. Endodontic microbiology. Ames, Iowa: Wiley-Blackwell; 2009, p. 68107.

56. Sogin ML, Morrison HG, Huber JA, Mark Welch D, Huse SM, Neal PR, et al. Microbial diversity in the deep sea and the underexplored rare biosphere. Proc Natl Acad Sci U S A 2006; 103: 1211520.

57. Teodoridis JM, Hall J, Marsh S, et al. CpG island methylation of DNA damage response genes in advanced ovarian cancer. Cancer Res. 2005; 65, 8961–8967.

58. Tost J, Dunker J, Gut IG. Analysis and quantification of multiple methylation variable positions in CpG islands by Pyrosequencing. Biotechniques. 2003; 35, 152–156.

59. Trama JP, Mordechai E, Adelson ME. Detection of Aspergillus fumigatus and a mutation that confers reduced susceptibility to itraconazole and posaconazole by real-time PCR and pyrosequencing. J. Clin. Microbiol. 2005a; 43: 906 - 908.

60. Trama JP, Mordechai E, Adelson ME. Detection and identification of Candida species associated with Candida Vaginitis by real-time PCR and pyro sequencing. Mol. Cell Probes. 2005b; 19(2): 145-152.

61. Tsibris A, Russ C, Paredes R, Arnaut R, Honan T, Cahill P, et al. Detection and Quantification of Minority HIV-1 ENV V3 Loop Sequences by Ultra-Deep Sequencing: Preliminary Results. 15th International HIV Drug Resistance Workshop; International Medical Press; 2006.

62. Turnbaugh PJ, Ley RE, Mahowald MA, VM, Mardis ER, Gordon

JI. An obesity-associated gut microbiome with increased capacity for energy harvest. Nature. 2006; 444:1027–1031.

63. Uhlmann K, Brinckmann A, Toliat MR, Ritter H, N'rnberg P. Evaluation of a potential epigenetic biomarker by quantitative methyl-single nucleotide polymorphism analysis. Electrophoresis. 2002; 23: 4072-4079.

64. Van Doorn LJ, Kleter B, Quint WG. Molecular detection and genotyping of human papillomavirus. Expert Rev. Mol. Diagn. 2001; 1: 394-402.

65. Wang C, Mitsuya Y, Gharizadeh B, Ronaghi M, Shafer RW. Characterization of mutation spectra with ultra-deep pyrosequencing: application to HIV-1 drug resistance. Genome Res. 2007; 17:1195–1201.

66. Wu H, Khanna D, Park G, et al. Interaction between RANKL and HLADRB1 genotypes may contribute to younger age at onset of seropositive rheumatoid arthritis in an inception cohort. Arthritis Rheum. 2004; 50: 3093–3103.

CHAPTER 5

ADVANTAGES AND FUTURE OF PYROSEQUENCING

Shiv Sanjeevi Sripathi[1]; Prerna Pandey[2]

[1]Shiv Sanjeevi Sripathi completed his BSc in Biotechnology from Mumbai University and secured second rank in Mumbai University and first rank in his college: Kishinchand chellaram College. Shiv secured first rank in his college KET's V.G.Vaze College and has worked on a stem cell project at the Specialized Centre for Cell Based Therapy (SCCT), KEM Hospital at Mumbai, following which he was a Junior Research Fellow at Microbiology & Cell Biology Department, Indian Institute of Science, Bangalore from August 2009 to November 2011. The field of work involved cloning of cell wall genes and transcription factors in E. coli & M. smegmatis. This was followed by a stint as a Research and Development Executive at Transasia Biomedicals, Mumbai where he worked on patent writing and Immunoassays from 2011 till 2013. In 2013, he entered the teaching field where he has been tutoring biology and chemistry to the students of 11th and 12th grades along with English grammar and communication skills to people of all age groups. Shiv also loves helping people write scientific manuscripts.

[2]Dr Prerna Pandey, a biotechnologist with several years of wet lab research experience, worked at International Center for Genetic Engineering and Biotechnology, New Delhi. Her field of research in PhD included isolation and molecular characterization of geminiviruses, genome sequencing, gene annotation, and gene silencing using the RNA interference technology. She has also worked at Transasia Biomedicals and Advance Enzyme Technologies as a scientist. Prerna has published papers in peer-reviewed journals, and has submitted a number of annotated Geminiviral genome sequences in the GenBank, including two novel ones. She has also completed her editing and proof-

reading courses from Society for the promotion of Editors and Proof-readers, UK and now works as a freelance scientific editor and writer. When Prerna is not busy with her assignments, she loves to explore historical places.

INTRODUCTION

The advent of human genome sequence has been highly beneficial for research in genetics. Pyrosequencing is the distinctive and systematic technique used for precise genotyping (King, *et al.*, 2007) while analyzing polymorphism. The capability to analyze nucleic acid sequences is a novel invention for the meticulous learning of biological systems. It opened doors for research using sequence based DNA investigation for short sequences of DNA, with detailed classification of nucleic acids. The technique distributes needing labeled nucleotides, gel-electrophoresis and labeled primers (Ronaghi, 2001). Thus, this technique has proved to be extremely apt and permits the investigation of genetic variations that include single-nucleotide polymorphisms (SNP) (Milan, *et al.*, 200), DNA methylation, reviews imbalance in RNA alleles, short repeats and indels, gene copy number and is easily automated.

Pyrosequencing has successfully replaced techniques for analysis of DNA sequencing as it requires smaller sample size yielding higher output and cost cutting. Also, advancement in the technique like universal and multiplex primer applications have reduced the cost while giving improved output (Fakruddin, *et al.*, 2012). It is an important tool in the identification of viral (Adelson, *et al.*, 2005), fungal (Trama, *et al.*, 2005a) and bacterial (Grahn, *et al.*, 2003) typing. Pyrosequencing technique along with pre-programmed nucleotide addition order, SNP analysis and EST sequencing has been used to check clones and has yielded reliable and accurate results (Gharizadeh, 2003).

Pyrosequencing helps determine the allele percentage for the assessment of pooled samples and in the identification of indels, tri-allelic and short duplicate polymorphisms. The standardized control sequence in pyrosequencing offers internal quality control. This technique has resulted in efficient and rapid single-nucleotide polymorphism assessment which includes several clinically significant polymorphisms (Wu, *et al.*, 2010).

SUMMARY OF ADVANTAGES OF PYROSEQUENCING

Studies have shown that pyrosequencing is a rapid, reliable and sensitive tool which provides real-time sequence information and is highly beneficial in cancer research. Unlike PCR which can read only known genetic sequences, pyrosequencing reads the actual target sequence and detects known and unknown genetic variations occurring in all regions of DNA target where mutations can occur. Sanger sequencing for mutation and methylation analysis provides qualitative information and quantitative Sanger sequence is a lengthy, labor-intensive cloning of PCR products before sequencing.

Pyrosequencing, on the other hand, provides flexible, easy sequencing without the need for cloning clubbed up with high accuracy quantification. It facilitates the determination and analysis of intricate mutations in one single assay. It further characterizes the cancer mutations if they are multi-variable or contagious. It also detects unknown mutations easily. Pyrosequencing is also an important tool in the quantification of methylation in multiple and consecutive CpG sites. Scientists can easily and rapidly quantify SNPs enabling them to provide the essential genetic alterations in tumors along with identification of new tumor markers.

Recently, the technique is used extensively to study the KRAS gene. Cancer KRAS is an oncogene which has an important role in tumor formation and development. A permanently active KRAS protein is indicated in the mutation in KRAS gene. This type of activation is involved with several developmental phases of cancer which includes proliferation, abnormal cell growth, differentiation and increased metastasis and invasion. Mutant KRAS is found in 35-5% CRC (colorectal cancers), 72-90% of pancreatic cancers and about 50% in lung cancers. Research shows that Testing for KRAS mutation will help sort the mode and appropriate drugs for the treatment of above-mentioned cancers. It will help in the pharmacogenesis which is designing patient specific drugs for quicker healing in cancer patients.

Pyrosequencing generates sequence signals instantly to the downstream of the primer as compared to Sanger technique of sequencing where a reading gap of approximately 20-30 bases exists from the sequencing primer. Since it commences from the first base to the primer, it makes designing a primer more flexible in pyrosequencing. It has limitations of read-length compared to di-deoxy sequencing; yet, it is a rapid technique which is suitable for short stretch DNA sequences. Sanger sequencing takes approximately 4 hours for preparation while it takes about 15 minutes in pyrosequencing. The cost of reagent is low for short DNA stretch sequences. The order of arrangement of nucleotide is easily programmed. Also, any alterations in the pyrogram expose mutations, insertions or deletions (Gharizadeh, *et al.,* 2007). The technique is performed in real-time, making base callings and nucleotide incorporation observations continuously in a given sample. Pyrosequencing makes automation for extensive screenings possible (Gharizadeh, *et al.,* 2003a).

Table 5.1: Significant points of differences between Sanger and pyrosequencing highlighting the advantages of the latter

Sr. No.	Conventional sequencing	Sanger sequencing
Primer	A reading gap of approximately 20-30 bases	instantly to the downstream of the primer
Preparation time	4 hours	Close to 15 minutes
Principle	Chain termination	Sequencing by synthesis
Detection	Involves radioactivity/ fluorescence	Chemiluminescent detection
Results	Radiogram	Pyrogram developed
Automation		Facilitates automation
Gels	need for gels or polymers	avoids the need for labeled primers, labeled nucleotides and gel electrophoresis
Cost	Scaling up is difficult: costly for diploid genomes	The reagent costs are considerably lower
An additional advantage		Order of addition of nucleotides is permitted.

(Fakruddin and Abhijit Chowdhury, 2012).

SPECIFIC ADVANTAGES CONFERRED WITH PYROSEQUENCING

Microbial Genotyping and Identification

HPV genotyping

Human papillomaviruses (HPV) belong to the *Papovaviridae* family, consisting of a circular double stranded DNA. With the help of pyrosequencing, more than 100 types of HPV are analyzed and more than 30 types are known to cause an infection in the cervical mucosa. HPV are essential components in the technique as they contribute to the etiology of some malignant and benign abrasions in humans, eg in cervical cancer where they are further classified as high or low-risk genotypes based on their oncogenic potential (Gharizadeh, *et al.,* 2001).

Bacterial typing

Pyrosequencing proved to be an advanced technique for the identification of bacterial strains leading to close discrimination between closely connected microorganisms. Determining nucleic acid bases led to the analysis and diagnosis of the microorganisms based on molecular differentiation thus substituting the phenotyping with genotypic approaches. The DNA markers normally used for typing have both, the variable and conserved regions where a DNA primer is employed. 16S rRNA which forms the structural unit of ribosomal subunit has been used for the investigation of phylogenetic relationships between microorganisms (bacteria). Numerous features of the 16S rRNA gene make it an essential target in clinical identification (Patel, 2001). Its presence exclusively in bacteria makes it the universal target for identification of different strains of bacteria. Its function has remained unaltered for several years, suggesting sequence changes to more likely reflect random changes, which will alter the molecule's function. The 16S rRNA gene is adequately sufficient to hold statistically essential sequence information (Gharizadeh, *et al.,* 2003).

Sequencing the entire 16S gene of DNA for clinical diagnostics is not a practical approach as with techniques prior to pyrosequencing; it would be expensive and labor-intensive to conduct several sequencing reactions. Fortunately, the entire gene need not be sequenced in most bacteria to achieve species-level/detail. The region of highest sequence heterogeneity lies in the first 500 sequence bases at the 5'-end (Harris and Hartley, 2003). As 16S rRNA genes which are most common in the bacterial pathogens found in the body fluids are sequenced, PCR primers can amplify them based on the conservative nature of 16S rRNA genes (McCabe, *et al.,* 1995).

Pyrosequencing technique was implied in selected bacterial families that cause severe neonatal infections. Results showed that the technique can be further developed and can be applied to the various regions on various bacterial strains proving essential in the diagnosis clinical studies. Also, further advancement in pyrosequencing can be done to make the assay suitable for regions apart from 16s rRNA of the diverse species of bacteria (Gharizadeh, *et al.,* 2003).

Fungal pathogens

Fungi and yeasts are omnipresent, some of which are normal residents in the body existing as opportunistic organisms which can cause infections in the host in favorable conditions (Mannarelli and Kurtzman, 1998). They can cause an infection from acute to chronic as their incidences resulting in life threatening fungal infections is on a rise owing to an increased number of immunocompromised patients in recent years (Anaissie *et al.,* 1989). Resistance to drugs is a severe issue when it comes to treating yeast infections. Hence there is a need for accurate n rapid means to detect the infection-causing yeast as it will allow the delivery of the most efficient drug and its correct dosage.

Several experiments were performed obtaining samples from patients with invasive fungal infections. The samples were amplified through PCR techniques and analysis was done using gel electrophoresis, ELISA and pyrosequencing. The results of pyrosequencing method were in accordance with the respective cultural results making it an essential diagnostic tool for the identification and detection of invasive fungal strains. It also proved to be a quick, reliable and reproducible technology for the detection of fungal pathogens (Gharizadeh, *et al.,* 2004).

DNA amplification is essential for the process and it is the 18s rRNA gene which allows the amplification of a wide range of fungal species. The amplified fragments of DNA are sequenced to 40 bases for identifying the samples and this is carried out by pyrosequencing. The technique is reproducible and has been established by sequencing amplicons many times yielding identical sequence data every time for each sample. The data attained by pyrosequencing suggests 18-32 bases are enough to identify *Aspergillus spp., C. tropicalis, C. krusei, C. glabrata, C. albicans* (Gharizadeh et al., 2003b).

EST sequencing

Pyrosequencing when applied to EST sequencing followed by gene identification, proved to be accurate in determining the sequence information which was achieved by di-deoxy method (Nordstrom, *et al.*, 2001). Later, with the advancement of pyrosequencing automation and chemistry, increased levels of read-length sequences and screening of cDNA libraries is possible.

SNP analysis

Single nucleotide polymorphism (SNPs) is common and molecular form of gene variations (Collins, *et al.*, 1997) which are present in every 200-300 nucleotides on an average (Stephens, *et al.*,2001), meaning nearly 15 million SNPs present in the human genome (Salisbury, *et al.*, 2003). Each SNP symbolizes the difference in a particular nucleotide and can also replace itself with another nucleotide in a particular DNA stretch. They thus act as biological markers for extensive genome associated studies, aiding in the discovery of genes present in the common and complex diseases, helping locate the particular genes connected with a particular disease. Pyrosequencing technique when used on four SNPs located on 9q and 17p chromosomes from random subjects sequenced the samples accurately (Nordstrom, *et al.*, 2000b and Ahmadian, *et al.*, 2000). This remarkable feature helps in making SNP typing extremely easy and accurate which also suitable for high-throughput and automation analysis.

Tag Sequencing

Pyrosequencing is used to give results identical to those obtained by Sanger DNA sequencing technique. It proposed high-throughput analysis from human cDNA libraries as 96 samples can be investigated and analyzed in less than an hour. Since the original cDNA clone is available, pyrosequencing has an advantage in library screening for further investigations (Ronaghi, 2001).

Resequencing

An application of pyrosequencing for the field of resequencing offers an advantage in terms of longer read-length as nucleotide delivery is specified based on the order of the sequence. The technique generates an indication signal with every addition in a pyrogram; thus, any variation in the pattern symbolizes the presence of the mutation. This technique has been applied for resequencing p53 tumor suppressor gene and the resultant mutations were effectively quantified (Garcia, *et al.,* 2000)

EXAMPLES OF PYROSEQUENCING:

Case Study 1: bacterial communities in human gut

Wu, *et al.,* conducted studies on sampling methods to characterize the various bacterial communities residing in the human gut. They used 454/Roche pyrosequencing for the 16S rRNA gene of bacteria for the quantification of bacterial taxa (Wu *et al.,* 2010). Sequencing, amplification while comparing databases, allows identifying the bacterial lineage. Pyrosequencing helps in obtaining the data efficiently and economically.

Case Study 2: primers

Initially, pyrosequencing was restricted to sequence short stretches of DNA. The occurrence of an isomer of dATPαS; an alternate for natural dATP, was a limiting factor in the technique as it subdued the enzyme action in pyrosequencing chemistry. Removal of this non-functional nucleotide, Gharizadeh and team achieved DNA read-lengths up-to 100 bases (Gharizadeh, 2003). The introduction of polymerase Sequenase enabled the sequencing of homopolymericT-regions, enabling sequencing of such templates with harmonized extension which lead to a multitude of types or species or non-specific amplified products. The process was applied in various clinical infections and variants and it detected the causative agent in samples, eliminating the need for lengthy processes stringent PCR and cloning. The technique worked successfully with samples having subdominant species or types and low PCR yield samples, avoiding the redoing of the experiment for unproductive PCRs. Introducing a variety of genotype in a particular sample resulted in typing of multiple target DNA in the sample. Also, primers which are target-specific could be altered and modified as per the preferred application or clinical setup based on the prevalence of pathogens (Gharizadeh, *et al.*, 2004).

Case Study 3: formalin samples

Pyrosequencing technique has yielded quantitative data regarding the amount of methylation in multiple CpGs in the given sample. The technique used an inter conversion of a cytosine to a uracil nucleotide using amplification.

The technique proved to be useful in samples of DNA extracted from formalin-fixed in paraffin-embedded tissues (Doyle, *et al.*, 2011).

Case Study 4: Cancer

Two types of mutations are found in non-small-cell lung cancer (NSCLC) patients with 90% of epidermal growth factor receptor (EGFR) in lung adrenocarcinoma. Various techniques have been devised for the detection of these mutation causing genes but DNA sequencing providing 50% tumor cells stands out to be the best one (Gazdar, et al., 2009). Dufort and team performed experiments using pyrosequencing assay with the basis nested PCR to differentiate mutations on formalin-fixed in paraffin-embedded tumor tissues. They found pyrosequencing to be highly effective in providing accurate results in detecting the mutations in patients with NSCLC with a smaller sample size of 20% tumor cells (Dufort, *et al.*, 2011).

FUTURE TRENDS

Sequencing of genomes endows the various biological areas with immense information. Techniques with higher accuracy and efficiency for the detection of variations in genome scanning will be of great importance. It is a simple, straightforward, accurate, versatile and user-friendly technique having distinctive methodological characteristics which are appropriate for abundant applications in modern biology, outliving the Sanger's sequencing technique.

The technique is used in various fields like academics, industrial and clinics. It has shown high accuracy when used for the analysis of fragments of polymorphic DNA. The technique also yields quantification in the occurrence of alleles in populations. Currently, there is a need for rapid techniques which will also work with small sample size, high throughput and economical. Pyrosequencing has taken over various processes of DNA sequencing owing to all the mentioned reasons and its less preparation time, low reaction time and higher read-length along with large sample size pooled in a single pyrosequencing reaction (Gharizadeh et al., 2007).

There is scope for further improvement in the technique by improving chemistry, the efficiency of sequencing at higher temperatures (Ronaghi, 2001).

Removal of factors which inhibit the process will also lead to improved quality of the sequence and its read-length which will bring new insights to resequencing, microbial, fungal, viral typing and mutation detection.

Longer read-lengths will also enhance the time-frame for sequencing, decrease the sample size and enhancement in automation are areas that should be worked upon (Fakruddin, et al., 2012).

Achieving miniaturization of the technique in high-density microarrays, microtiter plates, or microfluidics will be highly economical. It can also increase the magnitude of orders which will lead to extensive genetic testing (Ronaghi, 2001).

Pyrosequencing does have challenges in the length of the reads as compared to the Gold standard: di-deoxy sequencing. However, the speed offered by the technique and its results in terms of real time all facilitate its use for short pieces of DNA (Gharizadeh et al., 2007).

454 Life Sciences Corporation (Branford, CT, USA) developed a platform that can sequence 300,000 samples within a time of 4 hours and an accuracy of 99% with an output of to 20- 40 million bases. The instrument integrates emulsion PCR along with sequencing of DNA

software that is based on microfluidic pyrosequencing (Gharizadeh *et al.*, 2007).

The technique of Pyrosequencing uses enzymes that work in a co-operative pattern: such enzymes require characterization of several factors such as K_M, kcat as well as sensitivity and specificity of the enzymes that can all be adjusted for optimal performance. The real-time monitoring of the enzymes is also permitted (Gharizadeh *et al.*, 2003).

As the technique of pyrosequencing, is recent; there is scope of for advancements in the instruments and the enzymes involved that adds to the versatility of the technique. The procedures offers an advantage of cost as well as time so further work can result in further improvements in technology.

Pyrosequencing also holds a great potential in Cancer Research. Each year a lot of funds are invested on cancer prescription which later turns out to be hazardous to the patients. Studies show that the prescribed drugs are barely effective for 60% of patients and most deaths happen due to incorrectly prescribed drugs. This is a result of the inability to target the right mode of treatment to patients which leads to frequent deaths and also an elevation in treatment cost. Studies on KRAS gene is a potential tool to offer personalized treatment to patients, though it is in its earlier phase. With an advent in more research using pyrosequencing, maximum safety and efficacy for individual patients and specific groups is a possibility.

CONCLUSION

The technique of pyrosequencing offers several advantages that include increased sensitivity as well as a facility of automation. This chapter covered the advantages of pyrosequencing with specific merits conferred using examples. Case studies were also included to further reinforce the advantages with this technique.

With its initial restriction to sequencing only short stretches of DNA, methodological improvements have aided the reading of longer sequences of DNA. With further development in its chemistry and instrumentation, the technique is poised for a giant leap with further advantages.

REFERENCES

1. Adelson ME, Feola M, Trama J, Tilton RC and Mordechai E. Simultaneous detection of herpes simplex virus types 1 and 2 by realtime PCR and pyrosequencing. J. Clin. Virol. 2005; 33:25-34.
2. Ahmadian A, Gharizadeh B, Gustafsson AC, Sterky F, Nyren P, Uhlen M, Lundeberg J. Single-nucleotide polymorphism analysis by pyrosequencing. Anal. Biochem. 2000; 280:103-110.
3. Anaissie EJ, Bodey GP and Rinaldi MG. Emerging fungal pathogens. Eur. J. Clin. Microbiol. Infect. Dis. 1989; 8:323-330.
4. Collins FS, Guyer MS, Charkravarti A. Variations on a theme: cataloging human DNA sequence variation. Science. 1997; 278:15801581.
5. Doyle B, O'Riain C, Appleton K. Pyrosequencing of DNA extracted from formalin-fixed paraffin-embedded tissue. Methods Mol. Biol. 2011; 724:181-190.
6. Dufort S, Richard MJ, Lantuejoul S, Fraipont F. Pyrosequencing, a method approved to detect the two major EGFR mutations for anti EGFR therapy in NSCLC. J. Exp. Clin. Cancer Res. 2011; 30(1):57.
7. Fakruddin and Abhijit Chowdhury. Pyrosequencing-An Alternative to Traditional Sanger Sequencing. Am. J. Biochem. Biotechnol. 2012; 8 (1): 14-20.
8. Fakruddin MD, Chowdhury A, Hossain MDN, Mazumdar RM, Shahnewaj MD, Reaz, et al. Pyrosequencing Principles and applications. International Journal of Life Science and Pharma Research 2012; 2:65-76.
9. Garcia AC, Ahamdian A, Gharizadeh B, Lundeberg J, Ronaghi M, Nyren P. Mutation detection by Pyrosequencing: Sequencing of exons 5 to 8 of the p53 tumour supressor gene. Gene. 2000; 253:249– 257.
10. Gazdar AF. Activating and resistance mutations of EGFR in nonsmall-cell lung cancer: role in clinical response to EGFR tyrosine kinase inhibitors. Oncogene. 2009; 28(Suppl 1):S24–31.
11. Gharizadeh B, Ghaderi M, Nyren P. Pyrosequencing technology for short DNA sequencing and whole genome sequencing. Technology. 2007; 47:129-132.

12. Gharizadeh B, Kalantari M, Garcia CA, Johansson B and Nyren P Typing of human papillomavirus by pyrosequencing. Lab Invest. 2001; 81:673-679.
13. Gharizadeh B, Kalantari M, Garcia CA, Johansson B, Nyren P. Typing of human papillomavirus by pyrosequencing. Lab Invest 2001; 81:673-679.
14. Gharizadeh B, Norberg E, Löffler J, Jalal S, Tollemar J, Einsele H, Klingspor L, Nyrén P. Identification of medically important fungi by the Pyrosequencing technology. Mycoses. 2004; 47(1-2):2933.
15. Gharizadeh B, Ohlin A, Molling P, Backman A, Amini B, Olcen P, Nyren P. Multiple group-specific sequencing primers for reliable and rapid DNA sequencing. Mol. Cell Probes. 2003; 17:203210.
16. Gharizadeh B. Method development and applications of Pyrosequencing technology. Doctoral dissertation from the Department of Biotechnology, Royal Institute of Technology, Stockholm, Sweden (2003).
17. Grahn N, Olofsson M, Ellnebo-Svedlund K, Monstein HJ and Jonasson J. Identification of mixed bacterial DNA contamination in broadrange PCR amplification of 16S rDNA V1 and V3 variable regions by pyrosequencing of cloned amplicons. FEMS Microbiol. Lett. 2003; 219:87–91.
18. Harris KA, Hartley JC. Development of broad-range 16S rDNA PCR for use in the routine diagnostic clinical microbiology service. J. Med. Microbiol. 2003; 52:685-91.
19. King CR, Scott-Horton T. Pyrosequencing: a simple method for accurate genotyping. Methods Mol. Biol. 2007; 373:39-56.
20. Mannarelli BM, Kurtzman CP. Rapid identification of Candida albicans and other human pathogenic yeasts by using short oligonucleotides in a PCR. J. Clin. Microbiol. 1998; 36:1634-1641.
21. Marsh S. Pyrosequencing applications. Methods Mol. Biol. 2007; 373:15-24.
22. McCabe KM, Khan G, Zhang YH, Mason EO and McCabe ER. Amplification of bacterial DNA using highly conserved sequences: automated analysis and potential for molecular triage of sepsis.

Pediatrics. 1995; 95:165-169.

23. Milan D, Jeon JT, Looft C, Amarger V and Robic A. A mutation in PRKAG3 associated with excess glycogen content in pig skeletal muscle. Science. 2000; 288:1248-1251.
24. Nordstrom T, Gharizadeh B, Pourmand N, Nyren P, Ronaghi M. Method enabling fast partial sequencing of cDNA clones. Anal. Biochem. 2001; 292:266-271.
25. Nordstrom T, Ronaghi M, Forsberg L, de Faire U, Morgenstern R and Nyren P. Direct analysis of single-nucleotide polymorphism on double-stranded DNA by pyrosequencing. Biotechnol. Appl. Biochem. 2000b;31(2):107-112.
26. Patel JB. 16S rRNA gene sequencing for bacterial pathogen identification in the clinical laboratory. Mol. Diagn. 2001; 6:313-321.
27. Ronaghi M, Nygren M, Lundeberg J, Nyren P. Analyses of secondary structures in DNA by pyrosequencing. Anal. Biochem. 1999; 267:65-71.
28. Ronaghi M. Improved Performance of Pyrosequencing Using Single-Stranded DNA-Binding Protein. Anal. Biochem. 2000; 286(2):282-288.
29. Ronaghi M. Pyrosequencing sheds light on DNA sequencing. Genome Res. 2001; 11(1):3-11
30. Salisbury BA, Pungliya M, Choi JY, Jiang R, Sun XJ and Stephens JC. SNP and haplotype variation in the human genome. Mutat. Res. 2003;526:53-61.
31. Stephens JC, Schneider JA, Tanguay DA, Choi J, Acharya T, Stanley SE, et al. Haplotype variation and linkage disequilibrium in 313 human genes. Science. 2001; 293:489-93.
32. Trama JP, Mordechai E and Adelson ME. Detection of Aspergillus fumigatus and a mutation that confers reduced susceptibility to itraconazole and posaconazole by real-time PCR and pyrosequencing. J. Clin. Microbiol. 2005a; 43: 906 - 908.

33. Wu GD, Lewis JD, Hoffmann C, Chen YY, Knight R, Bittinger K, et al. Sampling and pyrosequencing methods for characterizing bacterial communities in the human gut using 16S sequence tags. BMC Microbiol. 2010; 10:206

CHAPTER 6

THE USE OF GENUS-SPECIFIC AMPLICON PYROSEQUENCING TO ASSESS PHYTOPHTHORA SPECIES DIVERSITY USING EDNA FROM SOIL AND WATER IN NORTHERN SPAIN

Santiago Català[1], Ana Pérez-Sierra[1,2], and Paloma Abad-Campos[1]

[1]Instituto Agroforestal Mediterráneo, Universitat Politècnica de València, Camino de Vera s/n, Valencia, Spain,

[2]Forest Research, Alice Holt Lodge, Farnham, Surrey, United Kingdom

ABSTRACT

Phytophthora is one of the most important and aggressive plant pathogenic genera in agriculture and forestry. Early detection and identification of its pathways of infection and spread are of high importance to minimize the threat they pose to natural ecosystems. eDNA was extracted from soil and water from forests and plantations in the north of Spain. *Phytophthora*-specific primers were adapted for use in high-throughput Sequencing (HTS). Primers were tested in a control reaction containing eight *Phytophthora* species and applied to water and soil eDNA samples from northern Spain. Different score coverage threshold values were tested for optimal *Phytophthora* species separation in a custom-curated database and in the control reaction. Clustering at 99% was the optimal criteria to separate most of the *Phytophthora* species. Multiple Molecular Operational Taxonomic Units (MOTUs) corresponding to 36 distinct *Phytophthora* species were amplified in the environmental samples. Pyrosequencing of amplicons from soil samples revealed low *Phytophthora* diversity (13 species) in comparison with the 35 species detected in water samples. Thirteen of the MOTUs detected in rivers and streams showed no close match to sequences in international sequence databases, revealing that eDNA pyrosequencing is a useful strategy to assess *Phytophthora* species diversity in natural ecosystems.

INTRODUCTION

In recent years the increase of global plant trade and human movement have promoted the risk of introduction of invasive plants and exotic pathogens [1–3]. Biological invasions operate globally and are considered to be the second cause of biodiversity loss after direct habitat alteration and destruction. In this context, *Phytophthora* species are of particular importance worldwide as they are major pathogens in agriculture, horticulture and forestry causing important economic and ecological losses.

Environmental DNA (eDNA) from samples such as soil, water, air or permafrost is a complex mixture of genomic DNA from living cells and extracellular DNA from natural cell death from many different organisms [4]. In recent years, the examination of the eDNA in ecological studies has been used to characterize microbial and fungal communities using high-throughput sequencing (HTS) technologies [5–10] and to detect alien species in soil samples [11]. eDNA has also been used successfully to detect low density populations in freshwater environments [12], lakes [13], [14] and aquifers [15]. Most studies have focused on the detection of all organisms present in environmental samples using rDNA genes [16–19] but only a few have targeted only one organism/genera [20], [21].

Traditionally, different methods have been used for the detection of *Phytophthora* species based on isolation on selective media or using different baiting techniques. However, these techniques are time-consuming, and sometimes produce false negatives due to the low inoculum available. Furthermore, identification based on morphological characteristics requires specific taxonomical expertise and a considerable effort. In contrast, different molecular-based methods have been developed and applied for the detection of *Phytophthora* spp. in environmental samples. These methods allow fast and accurate pathogen detection and identification even when the inoculum amount is low, and quantitation of specific *Phytophthora*pathogens is also possible using Real-Time PCR assays [22–25]. A membrane-based oligonucleotide array was developed for the detection of 98 described and 15 undescribed species of *Phytophthora* [26]. Cloning was also applied to assess the *Phytophthora*communities present in soil and water samples [27], but the low throughput of the technique lowers the probability of detecting rare or newly introduced species with low inoculum levels. Different studies have applied HTS for the detection of *Phytophthora* species in soil samples [10], [11], [28], [29], but not, to date, for water.

Increasing economic and environmental losses caused by invasive *Phytophthora* species in natural ecosystems justifies the implementation of an efficient and rapid technique for their detection and accurate identification. In this study, genus-specific primers were adapted to assess *Phytophthora* species diversity in natural ecosystems using high-throughput amplicon pyrosequening of eDNA from soil and water environments, based in the polymorphic and widely accepted barcoding target Internal Transcribed Spacer 1 (ITS1), and validated with a control reaction with DNA of pure cultures. The objective of this study was to apply HTS to investigate the presence of *Phytophthora* in different plant communities in natural forests, plantations and aquatic environments in the north of Spain.

MATERIAL AND METHODS

Culture samples

Pure cultures of *Phytophthora gonapodyides* (PS-1512), *P.* taxon PgChlamydo (PS-1510), *P. lacustris* (PS-1513), *P. cryptogea* (PS-1584), *P. citrophthora* (PS-1544), *P. plurivora* (PS-1514), *P cambivora*

(PS-1556) and *P. cinnamomi* (PS-1520) were obtained from the fungal culture collection of the Instituto Agroforestal Mediterráneo, Universitat Politècnica de València (Spain). The identities of these cultures were previously determined by sequencing of ITS region.

Sampling areas and environmental samples

Two geographical areas were studied. Villanúa (Central Pyrenees, Aragón, Spain) which was selected as a pure European silver fir (*Abies alba*) forest (42°04′13″N, 0°03′13″W) and Irati Forest (42°56′-43°00'N, 1°10′-0°58'E) located in northern Spain (Western Pyrenees, Navarra, Spain). The main vegetation type of Irati Forest is either native beech stands (*Fagus sylvatica*) or mixed beech and European silver fir (*Abies alba*) forest, although it is possible to find recent plantings of Lawson Cypress (*Chamaecyparis lawsoniana*) or Douglas fir (*Pseudotsuga menziesii*) in some areas.

Special authorization for sampling in the protected area of Irati Forest was granted by the Administración Forestal del Gobierno de Navarra. No special permit was required for sampling in Villanúa. This study did not involve endangered or protected species.

Soil samples

In Villanúa six soils from beneath European silver fir (AS) were collected in June 2012. In Irati Forest, soil samples were collected in October 2012 from 24 different locations selected according to their main vegetation community. The number of samples per vegetation type was proportional to their area in the forest: ten samples were collected from beech stands (F), six from Douglas fir (PS), five from European silver fir (AB), and three from Lawson Cypress (CH). Each sample consisted of soil subsamples from six different points selected at random that were mixed together (approx. 3 kg of soil/sample). Samples were collected around the trees at 1 meter distance from the main trunk by digging to a depth of about 30 cm.

Water samples

In total 15 water samples were analyzed, two water samples from Villanúa (R1) and 13 water samples from rivers and streams in Irati Forest (R2) during October 2012. Samples were taken after several days of rain. Water samples (10 L) were collected from rivers and streams using plastic containers. The water was filtered *in situ* using a clean knap-sack sprayer to pump the water fitted with a 47 mm polypropylene Swinnex filter holder (Millipore Corp. Bedford, MA, USA) with a 5 μm pore-size autoclaved filter (cellulose acetate and cellulose nitrate mixture) (SMWP04700, Millipore, Ireland). Initially, the collected water was filtered into the sprayer through a 100 μm pore size mesh to remove any debris. Then the water was pumped using three filters per 10 L of water. Each filter was carefully removed with a pair of sterilised tweezers and cut in half with disinfested scissors. The filter was kept in a sterilized 15 ml polypropylene tube (Deltalab S.A., Barcelona, Spain). The tubes were kept in a cool-box during field sampling and in the freezer (-20°C) once in the laboratory. The sprayer was washed with 70% ethanol and rinsed with water thoroughly before sampling in each location.

DNA extraction

Each soil sample (up to 3 kg) was well mixed and homogenized by sieving (2 mm mesh size), and 50–80 g was lyophilized overnight and then crushed using FRITSCH Variable Speed Rotor Mill-PULVERISETTE 14 (ROSH, Oberstein, Germany). Samples were maintained at 5°C until DNA extraction.

Up to three subsamples were taken from each soil sample, and total genomic DNA was extracted from 300 mg of each subsample using the ZR Soil Microbe DNA MiniPrep (Zymo Research, Irvine, USA) following the manufacturer's instructions but with the final elutions in 50 μL of elution buffer instead of in a 100 μL of elution buffer.

Filters from water samples (3 filters/sample) were first frozen at −80°C and then disrupted using the TissueLyser LT compact bead mill (Qiagen, UK). DNA was extracted from each of the disrupted filters using the E.Z.N.A. Plant Kit (Omega Bio-tek, Doraville, USA) according to the manufacturer's recommendations but with the final elutions in 50 μL of elution buffer instead of in a 100 μL of elution buffer.

DNA was extracted from pure cultures using the E.Z.N.A. Plant Kit (Omega Bio-tek, Doraville, USA) according to the manufacturer's recommendations.

Control reaction

Each of the genomic DNA from cultures was diluted 10, 100 and 1000 times and amplified separately using a SYBR green real-time PCR assay with the *Phytophthora*-specific primers 18Ph2F and 5.8S-1R [27]. DNAs presenting similar Cycle Threshold values were mixed together to account for differences in ITS copy number.

Amplicon library generation and 454-pyrosequencing

Amplicon libraries were generated using a nested PCR approach using a Hot Start polymerase (Dominion MBL, Córdoba, Spain). In the first PCR round the *Phytophthora*-specific primers 18Ph2F and 5.8S-1R were used [27]. For the second PCR round, fusion primers were designed following the GS Junior System Guidelines for Amplicon Experimental Design [30], selecting the unidirectional sequencing protocol for library construction (Lib-L chemistry for emulsion PCR, emPCR, 'One-Way Reads'). The template-specific sequence of the forward fusion primer was the universal ITS6 primer [31] (5'-A-KEY-MID-ITS6–3'), while that of the reverse fusion primer was the same reverse primer used in the first PCR round (5'-B-KEY-5.8S-1R-3'), where A and B represent the pyrosequencing adaptors and the multiplex identifier (MID) was added for postsequencing sample identification. In the case of soil samples, 1 μL of genomic DNA was used in the first PCR round and in the case of water samples the genomic DNA was diluted ten times (approximately 0.2–2 ng was used as template in both DNA sources). The PCR products from the first round were diluted 10 times (soil samples) or 100 times (water samples) for the second PCR round, and in this case, 1 μL was used as a template. The PCR conditions were: 1 cycle of 95°C for 2 min, 30 cycles (1st round) or 25 cycles (2nd round) of 95°C for 20 s, 60°C for 30 s, 72°C for 30 s and a final extension of 72°C for 7 min. For the control reaction, 1 μl of genomic DNA was used in the first PCR round (30 cycles), and 1 μl of PCR product was used as template in the second round (25 cycles).

The PCR products were visualized in a Bioanalyzer 2100 (Agilent Technologies, Palo Alto, CA, USA). The amplicons from soil and pure culture samples were double purified using the Agencourt AMPure XP Bead PCR Purification protocol (Beckman Coulter Genomics, MA,

USA). However, in the case of water samples a different method was used with the aim to avoid non-specific products corresponding with other organisms with longer ITS1 fragments than *Phytophthora*, as well as chimeras, which could reduce the quality of the sequencing. Each PCR product from water samples was cleaned using E-Gel SizeSelect (Invitrogen, Burlington, ON, Canada) and the bands between 290 and 450 bp were recovered for sequencing.

After purification, the amplicons were visualized in a Bioanalyzer 2100 and quantified by fluorimetry using Quant-iT PicoGreen kit (Invitrogen Molecular Probes, Eugene, Oregon, USA). The three PCR products from each soil and water samples were mixed together based on their concentration to obtain a single product. The 45 resultant amplicons from the seven areas (each area identified with an individual MID) were pooled at equal concentrations for sequencing.

Amplicon libraries were sequenced in a GS Junior 454 system (Roche 454 Life Sciences, Branford, CT, USA) by the Sequencing and Genotyping Service from the University of Valencia (Burjassot, Spain). Three different emPCR conditions were tested in three different sequencing runs. The first run was performed following Roche's standard protocol for emPCR amplification for the 'One-Way Reads' experimental design, the second run was using the same protocol but removing the larger amplicons by E-Gel SizeSelect, and in the third run, the emulsion was performed following the short length amplicon libraries protocol. Library from pure cultures was sequenced in a fourth sequencing run following the short length protocol.

Trimming and Molecular Operational Taxonomic Unit (MOTU) clustering

The sequences were first sorted into separate files according to their multiplex identifier (MID) using the sfffile script included in the Roche Newbler package (http://www.454.com/products/analysis-software/). Sequences from each SFF file were extracted using the sff_extract script (http://bioinf.comav.upv.es/sff_extract/index.html) to generate single FASTA, XML and quality files. FASTA and quality files were combined into a single FASTQ file using a Python script (Python version 2.7, Python Software Fdn) and then opened in FastQC software [32] to examine the length and quality of reads.

Reads were trimmed based on quality scores using Lucy software [33]. Sequence length shorter than 100 bp and low quality reads (average read quality below 20) were not considered for analysis. Sequences from the three sequencing runs were combined in a single FASTA file after quality trimming. A new Python script was created to trim the 3' end (B adaptor, key and reverse primer) and the 5' end (key, MID and forward primer). This script included a new step to remove sequences shorter than 100 bp after the second trimming.

Different score coverage threshold values were tested for MOTU clustering with a custom-curated database including 146 ITS1 sequences of described and new *Phytophthora* taxa, and the sequences of species used in the control reaction. The resultant FASTA file from each library was clustered with a length coverage threshold of 90% and a score coverage threshold of 99% (-L 0.9 -S 99) using blastclust software [34]. After clustering, a custom Python script was used to convert each MOTU from the output clustering list file into an individual FASTA file, sorting MOTUs in decreasing order of abundance, with a step to remove unique sequences (singletons), and then were aligned by MUSCLE [35]. Each alignment was manually checked in Seaview software [36], and finally, a consensus sequence was generated from each MOTU with the aim to reduce homopolymer and sequencing errors.

The consensus sequences of the MOTUs were identified using the BLAST tool in the GenBank database [34], the *Phytophthora* Database [37] and a custom-curated database comprising Sanger sequences showing a non-mixed chomatogram obtained from environmental samples and culture collections maintained at the Instituto Agroforestal Mediterráneo, Universitat Politècnica de València (Spain). A single sequence was selected at random in those MOTUs that had two reads. Consensus sequences were subjected to phylogenetic analysis to confirm the results obtained from BLAST searches.

RESULTS

Sequencing throughput and quality control

The three different emPCR conditions tested produced 1,142 good quality reads in the first run, 14,289 (10,233 used in the present study) in the second and 140,767 reads in the third run.

The total data set from the three sequencing runs comprised 152,142 good quality sequences. The results obtained for each library were: 7,206 sequences for library 1 (R1), 47,419 for library 2 (R2), 13,543 for library 3 (CH), 19,796 for library 4 (PS), 21,982 for library 5 (AB), 26,527 for library 6 (F) and 15,669 for library 7 (AS). After trimming 151,311 sequences were considered for analysis. The average quality of the filtered sequences was of 36 and the average read length was 306 bp.

Score coverage threshold analysis

A reference database comprising 146 ITS1 sequences of described and new *Phytophthora* taxa was used to test the optimal threshold value. A FASTA file containing the ITS1 sequences was created and clustered at different threshold values. Clustering at 100%, or at 99.5%, separated the higher number of *Phytophthora* species (S1 Fig.). Furthermore, based on ITS1 sequences it was not possible to separate 10% of the species using 100% score coverage threshold.

Using 99% coverage threshold allowed the separation of the highest number of *Phytophthora*species in the control reaction as shown in S2 Fig. Clustering at 99.5 or 100% also separated the eight *Phytophthora* species. However, the number of singletons increased exponentially due to the presence of sequencing and homopolymer errors, producing a high loss of data. Therefore, for all the analysis a 99% of score coverage threshold was used for MOTU clustering.

Control reaction

A total of 6,698 sequences were generated from the control reaction comprising eight *Phytophthora* species. After quality control, 6,683 sequences were considered for analysis. Applying the 99% score coverage threshold previously calculated a total of 555 MOTUs, including 529 singletons, were obtained. The MOTUs with highest number of sequences corresponded to the eight *Phytophthora* species. The identity of the consensus sequences with Sanger reads was the 100% or 99.5% in all cases. Mismatches were, in all cases, based on homopolymeric regions. The read distribution from each species is shown in Table 1.

Table 1: BLAST results of the clustered sequences from the control reaction applying a barcoding threshold value of 99%

MOTU	Species	Number of Reads	ID with Sanger reads (%)	Error
1	***P. gonapodyides***	**1190**	**100**	**-**
2	***P. cryptogea***	**1025**	**100**	**-**
3	***P. cambivora***	**980**	**100**	**-**
4	***P. taxon PgChlamydo***	**884**	**100**	**-**
5	***P. colocasiae***	**879**	**99.5**	**PolyT**
6	***P. lacustris***	**604**	**99.5**	**PolyT**
7	***P. cinnamomi***	**434**	**100**	**-**
8	***P. plurivora***	**121**	**99.5**	**PolyT**
9	*P. gonapodyides*	3	98	PolyA, -/T, -/G, -/C
10	Chimera	2	-	-
11	*P.* taxon *PgChlamydo*	2	96	A/-, PolyA, T/-, PolyG, T/-, T/-, G/-, -/T, -/T
12	*P. gonapodyides*	2	98	PolyA, -/G, C/T, -/G
13	Chimera	2	-	-
14	*P. gonapodyides*	2	98	G/A, T/-, G/-, T/-, G/-
15	*P. cambivora*	2	97	C/T, T/C, T/A, T/A, C/T, T/A
16	*P. gonapodyides*	2	98	PolyA, A/-, PolyT, PolyA
17	*P. gonapodyides*	2	97	PolyG, T/C, C/T, T/-, T/-, G/A
18	*P. gonapodyides*	2	98	C/T, A/G, T/C, T/C
19	*P.* taxon *PgChlamydo*	2	97	A/-, -/A, T/C x3
20	*P. colocasiae*	2	98	PolyA, G/-, PolyT x2
21	*P. cryptogea*	2	99	G/A, G/T
22	*P. colocasiae*	2	98	PolyA, G/-, T/-
23	*P. colocasiae*	2	96	A/-, PolyA, T/A, A/C,-/T, PolyT
24	*P. plurivora*	2	98	PolyT, T/C, A/T
25	*P. plurivora*	2	97	PolyA, PolyT, T/C, A/T
26	*P. plurivora*	2	98	PolyT x3

Results only include MOTUs with more than two sequences. Identity and mismatches with Sanger reads are also indicated. MOTUs with higher number of sequences are in bold.

doi:10.1371/journal.pone.0119311.t001

Analysis of MOTUs from soil eDNA samples

A total of 96,895 reads which passed quality control from soil samples (sampling areas CH, PS, AB, F and AS) were considered for analyses. Clustering of 96,895 reads resulted in 8,151 MOTUs, including 8,114 singletons, which were discarded for the analysis. The 37 nonsingleton MOTUs corresponded to 13 *Phytophthora* species, included in clades 1, 4, 6, 7 and 8 (Fig. 1) from four plant communities: *A. alba*, *C. lawsoniana*, *P. menziesii* and *F. sylvatica*.

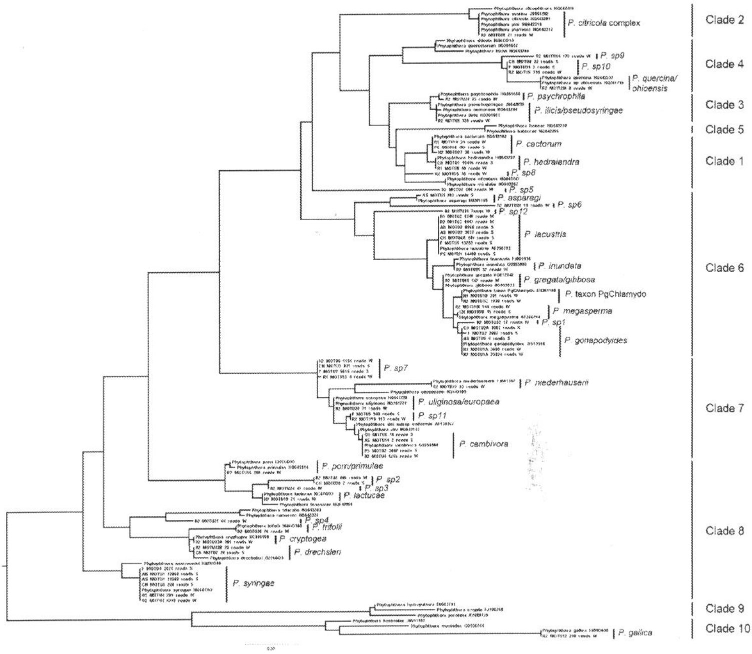

Figure 1: Unrooted phylogram based on nuclear ITS1 rDNA sequence analysis constructed with maximum likelihood approach. Each of the MOTU includes library precedence (R1, R2, CH, AB, F, PS or AS), number of MOTU resulted in the clustering at 99% of each library, number of reads and source (W, water; S, soil). Vertical bars indicate the *Phytophthora* species. MOTUs corresponding with undescribed species are indicated as "sp" (from sp1-sp12).

The specificity of the technique for the detection of *Phytophthora* in soils was 98.46%. The non-*Phytophthora* MOTUs (1.54% of nonsingleton clustered reads) matched with other plant pathogen genera such as *Pythium* and *Hyaloperonospora* (Table 2).

Table 2: Read distribution of species other from *Phytophthora* based on the ITS1 after BLASTn of the consensus sequences performed against GenBank. Reads from MOTUs with more than 99% similarity are in bold

	Villanua		Irati Forest					
Putative species detected non- *Phytophthora*	R1	AS	R2	CH	F	AB	PS	Total reads per species
Hyaloperonospora parasitica	**361**							361
Hyaloperonospora sp. 1							894	894
Hyaloperonospora sp. 2	248							248
Hyaloperonospora sp. 3			62					62
Hyaloperonospora sp. 4			117					117
Hyaloperonospora sp. 5	161							161
Hyaloperonospora sp. 6					198			198
Bremia lactucae			5					5
Peronospora aparines	**165**							165
Peronospora aestivalis			**232**					232
Peronospora glomerata			**150**					150
Pythium sp 1	7		2		6			15
Pythium sp 2		21						21
Pythiaceae 1	40							40
Pythiaceae 2	10							10
Pythium minus	**2**			**250**				252
Total reads per sampling area	994	21	568	250	204	0	894	

doi:10.1371/journal.pone.0119311.t002

Analysis of MOTUs from water eDNA samples

A total of 54,416 reads from water eDNA from the sampling areas R1 and R2 which passed quality control were considered for analyses. Clustering of 54,416 reads resulted in 3,646 MOTUs, including 3,548 singletons, which were discarded for the analysis. The 98 nonsingleton MOTUs corresponded to 35 *Phytophthora* species (Fig. 1) from clades 1, 2, 3, 4, 6, 7, 8 and 10 (Fig. 1). No *Phytophthora* species were detected from clades 5 and 9. The clades with the highest number of species were clade 6 and 8 with nine species in each. Only one species from clade 2 was found, matching with a *Phytophthora citricola*-complex species. The three most abundant MOTUs in water samples were *P. gonapodyides* (29,432 reads), followed by *P. lacustris* (6,341 reads) and *P. syringae* (5,528 reads). Fifteen sequences were obtained of a *Phytophthora* species (named as *Phytophthora* sp. 13) which sequences did not match any sequences in public databases and did not cluster in any of the ITS clades described. The closest match in GenBank was *Phytophthora* sp. REB326-69 (accession no. JX122744) with a sequence homology of 89%. This MOTU was not included in Fig. 1 as it was too divergent.

The specificity of the water eDNA pyrosequencing assay for *Phytophthora* detection was 96.92%. The non-*Phytophthora* MOTUs (3.08% of nonsingleton clustered reads) matched with other plant pathogen genera such as *Hyaloperonospora*, *Bremia*, *Peronospora*, *Pythium* or members of an undetermined Oomycete genus not represented in GenBank (Table 2).

DISCUSSION

The detection of *Phytophthora* from environmental samples using high-throughput amplicon pyrosequening of eDNA has shown high specificity and can be used hereafter to assess *Phytophthora* diversity in natural ecosystems.

Clustering at 100% was the best criteria to separate most of the species in the reference database. Other studies [38] adopted a barcoding threshold of 98% to assign most reads. However, in the current study we included a control species mixture and a 99% threshold was required to discriminate between closely related species resulting in the separation of 20% more species.

For the control reaction data, clustering at 99% of similarity was the minimum value able to separate all of the species with the minimum number of reads lost. Clustering at 99.5% or 100% resulted in the loss of reads due to the presence of homopolymer and sequencing errors that were discarded as singletons. Values below 99% were insufficient to separate *P. gonapodyides* and *P.* taxon PgChlamydo in separate MOTUs. In this study, clustering at 99% similarity allowed the separation of most species from clade 6 found in environmental water samples and reduced the risk of false MOTUs produced by sequencing errors. Although the pooling for the control reaction was made by mixing the DNAs based on similar CT values, the number of sequences obtained of *P. plurivora* was lower than expected and this could be a potential issue to consider when interpreting the results.

The limitations of the ITS1 region for *Phytophthora* taxonomic identification are particularly evident in clade 1, in clade 2 with the *P. citricola* complex [39] and in clade 6 which includes many aquatic species where many species are identical or differ by 1 bp in ITS1. Applying a similarity of 100% between sequences, it is still not possible to separate 10% of the species in the reference database, including species from clade 1 (*P. infestans*, *P. ipomoeae*, *P. mirabilis*, *P. andina*), from clade

2 (*P. capensis*, *P.* taxon emanzi) or clade 6 (*P. gibbosa*, *P. gregata*). Despite these limitations it remains a powerful tool for discrimination of most known species as well as the identification of new ones.

The conditions for emPCR amplification (2.2 million copies of template and 80 μL of Amp Primer A) used in the first and second sequencing runs resulted in light scattering into nearby wells and caused their elimination due to signal processing filtering. The third sequencing run was performed following guidelines for Short Length Libraries which reduced the amount of Amp Primer A in the Live Amp Mix from 80 to 20 μL. Although environmental ITS1 amplicons length did not match the Short Length Libraries definition (amplicons of 340 bp instead of amplicons <250 bp), the use of this protocol increased 100 fold the throughput in comparison with the first pyrosequencing run.

Similar studies based on pyrosequencing in soil samples from beech forests in France revealed the presence of only four *Pythiaceae* species (corresponding to the 0.8% of the sequences from the dataset), represented by two *Phytophthora* species: *P. plurivora* and an unidentified *Phytophthora* species from clade 7a [10]. Their methodology for library generation was based on a nested PCR approach, where the first PCR was based on the amplification of the whole ITS region with the primer pair ITS6/ITS4 [31], [40] and pyrosequencing was based on the ITS6/ITS7primers [31]. This pyrosequencing strategy revealed lower efficacy and specificity for *Phytophthora* detection in comparison with the methodology implemented in this study. The throughput of amplification of large targets, like whole ITS region in genus *Phytophthora* (up to 840 bp), is usually a problem in environmental samples, especially in soils, with abundant humic acids inhibitors and high DNA degradation. The importance of the length of the targeted region is that short DNA fragments (less than 300–400 bp) are usually very slowly degraded which allows their detection in environmental samples [41]. The implementation of large-scale eDNA-based ecological studies is highly dependent on the availability of suitable short metabarcodes [4]. Therefore, using in the first PCR the *Phytophthora*-specific primers as implemented, which include the ITS1 and a small portion of the 18S (amplicon length from 419 to 484 bp), rather than the whole ITS region provides better amplification results for metagenomics studies.

In contrast, other researchers [11] applied the same ITS6/ITS7 primers for pyrosequencing without using a nested approach, revealing the presence of 15 *Phytophthora* species in soils from chestnut forests in Italy. Their assay was very specific for *Phytophthora* species detection, where 78.8% of sequences (9,167 of 11,637 reads) matched *Phytophthora*.

In this study, a nested approach was used in order to yield suitable PCR products since no amplification was obtained with a single round PCR [27]. Nested PCR increases sensitivity and specificity compared to non-nested PCR techniques, but also increases the risk of cross-contamination events. To further reduce the occurrence of false-positives based on cross-contamination extremely care should be taken during DNA extraction, preparation of PCR reactions and post-PCR steps.

From a methodological point of view, soil pre-processing procedures (drying, sieving, lyophilizing, crushing and weighting) are time-consuming in comparison with the water filter processing. Working with water samples allows the processing and amplicon library generation in one day. Pyrosequencing of soil samples revealed lower *Phytophthora* diversity than from water samples, where only thirteen species were detected in five different ecosystems comprising 30 sampling sites. The highest number of species was detected in three sampling sites of *Chamaecyparis* plantations where 10 of the 13 *Phytophthora* species were detected. It is especially surprising that so few species were detected in *Abies alba* soil samples from Irati Forest, where only two species, *P. lacustris* and *P. syringae* were detected. Similarly within *Pseudotsuga menziesii* associated soil samples where only three species were detected. Pyrosequencing of water eDNA from rivers and streams revealed a hidden and unexpectedly high *Phytophthora* species diversity. A total of 35 *Phytophthora* species were detected, 13 of which could represent potentially novel species. *Phytophthora* species diversity from water samples in Irati Forest comprised all species detected in soil samples from the same sampling area. However, the species detected in water and soil samples in Villanúa were different. This could be due to the fact that a low number of samples were taken from this forest and also due to the different sampling times.

The DNA recovered from the filters is probably composed of DNA originated from living cells (mainly zoospores) and extracellular DNA originated from natural cell death in water. These could be present in the water itself or they could be have originated from washing off the soil after rain periods. Other studies [42] have demonstrated that DNA fragments in freshwater ecosystems persist for less than one month. This indicates that by sampling immediately after a rain event is possible to recover the maximum amount of DNA containing the potential *Phytophthora* community from the forests and water environments. This hypothesis is congruent with the presence of several *Phytophthora* species from most clades in water eDNA, including many plant pathogenic species usually isolated from roots such as *P. cambivora*, *P. niederhauserii*, *P. plurivora*, *P. pseudosyringae* or *P. syringae*.

Most of the *Phytophthora* species isolated from irrigation reservoirs and natural waterways belong to clade 6. These species, commonly called water moulds, show a strong association with both forest and riparian ecosystems, [43–46] and although some of them can be aggressive tree pathogens like *P. inundata* in Spain [47], it is hypothesized that they have a prevalently aquatic and saprophytic lifestyle [48]. In this study 73.6% of the reads belonged to clade 6, which may represent the resident community in rivers and streams. The most common species detected were *Phytophthora gonapodyides*, *P. megasperma*, *P.* taxon PgChlamydo, *P. inundata* and *P. lacustris*, which are commonly isolated by baiting from streams around the world [45], [49–53].

The *Phytophtora*-specific primers [27] used in the current study showed high specificity. In the original study the amplicons were cloned and a total of 260 amplicons were obtained with a complete absence of *Pythium s. lato* species. Only a minor cross reaction with two downy mildew phylotypes was found. However, in this study using the same primers with an HTS approach a total number of 152,142 reads were obtained with 97.9% specificity to *Phytophthora*. Only a small percentage (2.1%) of non-*Phytophthora* amplification was obtained, including the genus *Hyaloperonospora*, *Bremia*, *Peronospora* and *Pythium*.

Previous techniques (baiting of environmental samples and cloning) are time consuming and costly, and often are likely to generate false negatives due to the low frequency of some *Phytophthora* species and the inability of unknown Phytophthoras to grow on some artificial media. The technique implemented here overcomes these problems and can produce results in less than one week compared to several weeks or months by conventional techniques which would allow an early response to the threats posed by these pathogens. Although similar techniques have been used in previous studies it is the first time that genus-specific amplicon pyrosequencing has been used to target only *Phytophthora* in soil and water samples, unlocking previously hidden *Phytophthora* communities.

ACCESSION NUMBERS

The three pyrosequencing runs were combined into a single file and deposited in GenBank-SRA under the accession number SRP027499. Consensus nucleotide sequences of each *Phytophthora* MOTU were deposited in GenBank (accession no. AF132232 to AF132286).

SUPPORTING INFORMATION

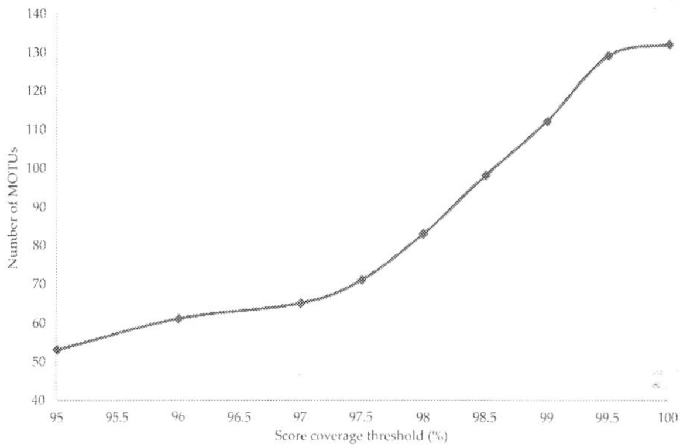

Figure S1: Number of MOTUs generated using different score coverage threshold values based on ITS1 using a reference database of Phytophthora species.

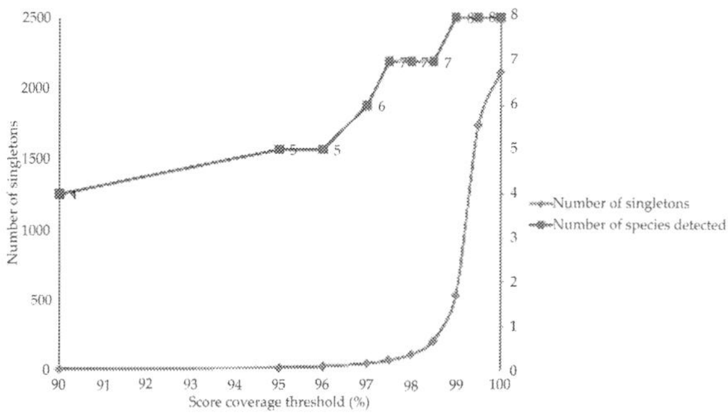

Figure S2: Number of *Phytophthora* species detected and singletons produced per score coverage threshold in the control reaction.

ACKNOWLEDGMENTS

We would like to thank the Departamento de Desarrollo Rural, Industria, Empleo y Medio Ambiente from Navarra for their support during the survey; Fernando Martínez-Alberola (University of Valencia) for his help with the bioinformatics procedures; the Sequencing Service from University of Valencia for their intensive work.

AUTHOR CONTRIBUTIONS

Conceived and designed the experiments: SC PAC APS. Performed the experiments: SC APS. Analyzed the data: SC. Contributed reagents/ materials/analysis tools: SC PAC APS. Wrote the paper: SC APS.

REFERENCES

1. Reichard SH, White P. Horticulture as a pathway of invasive plant introductions in the United States. Bioscience. 2001; 51: 103–113.
2. Jones DR, Baker RHA. Introductions of non-native plant pathogens into Great Britain, 1970–2004. Plant Pathol. 2007;56: 891–910.
3. Brasier CM. The biosecurity threat to the UK and global environment from international trade in plants. Plant Pathol. 2008; 57: 792–806.
4. Taberlet P, Coissac E, Hajibabaei M, Rieseberg LH. Environmental DNA. Mol Ecol. 2012;21: 1789–1793. pmid:22486819
5. Sogin ML, Morrison HG, Huber JA, Welch DM, Huse SM, Neal PR, et al. Microbial diversity in the deep sea and the underexplored "rare biosphere". Proc Natl Acad Sci USA. 2006;103(32): 12115–12120. pmid:16880384
6. Roesch LF, Fulthorpe RR, Riva A, Casella G, Hadwin A, Kent AD, et al. Pyrosequencing enumerates and contrasts soil microbial diversity. ISME Journal. 2007;1: 283–290. pmid:18043639
7. Acosta-Martínez V, Dowd S, Sun Y, Allen V. Tag-encoded pyrosequencing analysis of bacterial diversity in a single soil type as affected by management and land use. Soil Biol Biochem. 2008;40: 2762–2770.
8. Jumpponen A, Jones KL. Massively parallel 454 sequencing indicates hyperdiverse fungal communities in temperate *Quercus*

macrocarpa phyllosphere. New Phytol. 2009;184: 438–448. pmid:19674337

9. Nilsson RH, Ryberg M, Abarenkov K, Sjökvist E, Kristiansson E. The ITS region as a target for characterization of fungal communities using emerging sequencing technologies. FEMS Microbiol Lett. 2009;296(1): 97–101. pmid:19459974
10. Coince A, Caël O, Bach C, Lengell J, Cruaud C, Gavory F, et al. Below-ground fine-scale distribution and soil versus fine root detection of fungal and soil oomycete communities in a French beech forest. Fungal Ecol. 2013;6(3): 223–235.
11. Vannini A, Bruni N, Tomassini A, Franceschini S, Vettraino AM. Pyrosequencing of environmental soil samples reveals biodiversity of the *Phytophthora* resident community in chestnut forests. FEMS Microbiol Ecol. 2013;85(3): 433–442. pmid:23560715
12. Jerde CL, Mahon AR, Chadderton WL, Lodge DM. "Sight-unseen" detection of rare aquatic species using environmental DNA. Conserv Lett. 2011;4(2): 150–157.
13. Monchy S, Sanciu G, Jobard M, Rasconi S, Gerphagnon M, Chabé M, et al. Exploring and quantifying fungal diversity in freshwater lake ecosystems using rDNA cloning/sequencing and SSU tag pyrosequencing. Environ Microbiol. 2011;13(6): 1433–1453. pmid:21635672
14. Jobard M, Rasconi S, Solinhac L, Cauchie HM, Sime-Ngando T. Molecular and morphological diversity of fungi and the associated functions in three European nearby lakes. Environ Microbiol. 2012;14(9): 2480–2494. pmid:22568577
15. Livermore JA, Mattes TE. Phylogenetic detection of novel Cryptomycota in an Iowa (United States) aquifer and from previously collected marine and freshwater targeted high-throughput sequencing sets. Environ Microbiol. 2013;15(8): 2333–2341. pmid:23461624
16. Nakayama J, Jiang J, Watanabe K, Chen K, Ninxin H, Matsuda K, et al. Up to species-level community analysis of human gut microbiota by 16S rRNA amplicon pyrosequencing. Biosci Microbiota Food Health. 2013;32(2): 69–76. pmid:24936364
17. Creer S, Sinniger F. Cosmopolitanism of microbial eukaryotes in the

global deep seas. Mol Ecol. 2012;21(5): 1033–1035. pmid:22360453

18. Davey ML, Heegaard E, Halvorsen R, Kauserud H, Ohlson M. Amplicon-pyrosequencing-based detection of compositional shifts in bryophyte-associated fungal communities along an elevation gradient. Mol Ecol. 2013;22(2): 368–383. pmid:23190367
19. Weber CF, Vilgalys R, Kuske CR. Changes in fungal community composition in response to elevated atmospheric CO_2 and nitrogen fertilization varies with soil horizon. Front Microbiol. 2013;4.
20. Bergmark L, Poulsen PHB, Al-Soud WA, Norman A, Hansen LH, Sørensen SJ. Assessment of the specificity of Burkholderia and Pseudomonas qPCR assays for detection of these genera in soil using 454 pyrosequencing. FEMS Microbiol Lett. 2012;333(1): 77–84. pmid:22639954
21. Li L, Al-Soud WA, Bergmark L, Riber L, Hansen LH, Magid J, et al. Investigating the diversity of *Pseudomonas* spp. in soil using culture dependent and independent techniques. Curr Microbiol. 2013;67(4), 423–430. pmid:23677146
22. Schena L, Hughes KJ, Cooke DE. Detection and quantification of *Phytophthora ramorum*, *P. kernoviae*, *P. citricola* and *P. quercina* in symptomatic leaves by multiplex real-time PCR. Mol Plant Pathol. 2006;7(5): 365–379. pmid:20507453
23. Tooley PW, Martin FN, Carras MM, Frederick RD. Real-time fluorescent polymerase chain reaction detection of *Phytophthora ramorum* and *Phytophthora pseudosyringae*using mitochondrial gene regions. Phytopathology.2006;96(4): 336–345. pmid:18943415
24. Pavon CF, Babadoost M, Lambert KN. Quantification of *Phytophthora capsici* oospores in soil by sieving-centrifugation and real-time polymerase chain reaction. Plant Dis.2008;92(1): 143–149.
25. Than DJ, Hughes KJD, Boonhan N, Tomlinson JA, Woodhall JW, Bellgard SE, et al. A TaqMan real-time PCR assay for the detection of *Phytophthora* 'taxon Agathis' in soil, pathogen of Kauri in New Zealand. Forest Pathol. 2013;43(4): 324–330.
26. Chen W, Djama ZR, Coffey MD, Martin FN, Bilodeau GJ, Radmer L, et al. Membrane-based oligonucleotide array developed from multiple markers for the detection of many *Phytophthora* species.

Phytopathology. 2013;103(1): 43–54. pmid:23050746

27. Scibetta S, Schena L, Chimento A, Cacciola SO, Cooke DEL. A molecular method to assess *Phytophthora* diversity in environmental samples. J Microbiol Meth. 2012;88:356–368.
28. Català S, Pérez-Sierra A, Berbegal M, Abad-Campos P. First approach into the knowledge of the *Phytophthora* species diversity in Mediterranean holm oak forests based on 454 parallel amplicon pyrosequencing of soil samples. *Phytophthora* in Forest and Natural Ecosystems 6th International IUFRO Working Party 7.02.09 Meeting, Córdoba, Spain, pp 34; 2012.
29. Català S, Pérez-Sierra A, Beltrán A, Abad-Campos P. Next Generation Sequencing shows *Phytophthora* species diversity in soil samples of Macaronesian laurel forests from the Canary Islands. *Phytophthora* in Forest and Natural Ecosystems 6thInternational IUFRO Working Party 7.02.09 Meeting, Córdoba, Spain, pp. 86; 2012.
30. Roche. GS Junior System; Guidelines for Amplicon Experimental Design. Branford: 454 Life Sciences Corp. 45 p; 2010.
31. Cooke DEL, Drenth A, Duncan JM, Wagels G, Brasier CM. A molecular phylogeny of *Phytophthora* and related Oomycetes. Fungal Genet Biol. 2000;30:17–32. pmid:10955905
32. Andrews S. FastQC: a quality control tool for high throughput sequence data. Available: http://www.bioinformatics.bbsrc.ac.uk/projects/fastqc/. Accessed 2013 Feb 3.
33. Chou HH, Holmes MH. DNA sequence quality trimming and vector removal. Bioinformatics. 2001;17(12): 1093–1104. pmid:11751217
34. Altschul SF, Madden TL, Schäffer AA, Zhang J, Zhang Z, Miller W, et al. Gapped BLAST and PSI-BLAST: a new generation of protein database search programs. Nucleic Acids Res. 1997;25(17): 3389–3402. pmid:9254694
35. Edgar RC. MUSCLE: multiple sequence alignment with high accuracy and high throughput. Nucleic Acids Res. 2004;32(5): 1792–1797. pmid:15034147
36. Gouy M, Guindon S, Gascuel O. Seaview version 4: A multiplatform graphical user interface for sequence alignment and phylogenetic tree building. Mol Biol Evol. 2010;27(2): 221–224. pmid:19854763

37. Park J, Park B, Veeraraghavan N, Jung K, Lee YH, Blair JE, et al. *Phytophthora*database: a forensic database supporting the identification and monitoring of *Phytophthora*. Plant Dis. 2008;92(6): 966–972.
38. Vettraino AM, Bonants P, Tomassini A, Bruni N, Vannini A. Pyrosequencing as a tool for the detection of *Phytophthora* species: error rate and risk of false Molecular Operational Taxonomic Units. Lett Appl Microbiol. 2012;55: 390–396.
39. Jung T, Burgess TI. Re-evaluation of *Phytophthora citricola* isolates from multiple woody hosts in Europe and North America reveals a new species, *Phytophthora plurivora* sp. nov. Persoonia. 2009;22: 95–110. pmid:20198142
40. White TJ, Bruns TD, Lee SB, Taylor JW. Amplification and direct sequencing of fungal ribosomal RNA genes for phylogenetics. In: Innis MA, Gelfand DH, Sninsky JJ, White TJ, editors. PCR Protocols. San Diego: Academic Press, Inc. pp. 315–322; 1990.
41. Deagle B, Eveson JP, Jarman S. Quantification of damage in DNA recovered from highly degraded samples—a case study on DNA in faeces. Front Zool. 2006;3(1): 11.
42. Dejean T, Valentini A, Duparc A, Pellier-Cuit S, Pompanon F, Taberlet P, et al. Persistence of Environmental DNA in Freshwater Ecosystems. PLoS ONE. 2011; 6(8): e23398. pmid:21858099
43. Guha Roy S, Grunwald NJ. The plant destroyer genus *Phytophthora* in the 21st century. In book: Review of Plant Pathology, Edition: Volume 6, Publisher: Scientific Publishers (India), Jodhpur, Editors: B.N.Chakraborty, B.B.L.Thakore, pp. In press; 2014.
44. Brasier CM, Cooke DEL, Duncan JM, Hansen EM. Multiple new phenotypic taxa from trees and riparian ecosystems in *Phytophthora gonapodyides–P. megasperma* ITS Clade 6, which tend to be high-temperature tolerant and either inbreeding or sterile. Mycol Res. 2003;107: 277–290. pmid:12825496
45. Hüberli D, Hardy GESJ, White D, Williams N, Burgess TI. Fishing for *Phytophthora* from Western Australia's waterways: A distribution and diversity survey. Australas Plant Pathol. 2013;42: 251–260.
46. Jung T, Stukely MJC, Hardy GSJ, White D, Paap T, Dunstan W, et al. Multiple new *Phytophthora* species from ITS Clade 6 associated

with natural ecosystems in Australia: evolutionary and ecological implications. Persoonia. 2011;26: 13–39 pmid:22025801

47. Brasier CM, Sánchez-Hernández E, Kirk SA. *Phytophthora inundata* sp. nov., a part heterothallic pathogen of trees and shrubs in wet or flooded soils. Mycol Res. 2003;107: 477–484. pmid:12825521
48. Hansen EM, Reeser PW, Sutton W. *Phytophthora* beyond agriculture. Annu Rev Phytopathol. 2012;50:359–378. pmid:22681450
49. Reeser PW, Sutton W, Hansen EM, Remigi P, Adams GC. *Phytophthora* species in forest streams in Oregon and Alaska. Mycologia. 2011;103(1): 22–35. pmid:20943547
50. Reeser P, Sutton W, Hansen E. *Phytophthora* species in tanoak trees, canopy-drip, soil, and streams in the sudden oak death epidemic area of south-western Oregon, USA. New Zeal J For Sci. 2011;41: 65–73.
51. Nechwatal J, Bakonyi J, Cacciola SO, Cooke DEL, Jung T, Nagy ZA, et al. The morphology, behaviour and molecular phylogeny of *Phytophthora* taxon Salixsoil and its redesignation as *Phytophthora lacustris* sp. nov. Plant Pathol. 2012.62(2): 355–369.
52. Huai WX, Tian G, Hansen EM, Zhao WX, Goheen EM, Grünwald NJ, et al. Identification of *Phytophthora* species baited and isolated from forest soil and streams in northwestern Yunnan province, China. Forest Pathol. 2013;43(2): 87–103.
53. Oh E, Gryzenhout M, Wingfield BD, Wingfield MJ, Burgess TI. Surveys of soil and water reveal a goldmine of *Phytophthora* diversity in South African natural ecosystems. IMA Fungus. 2013;4(1): 123–131. pmid:23898418

CHAPTER 7

454-PYROSEQUENCING REVEALS VARIABLE FUNGAL DIVERSITY ACROSS FARMING SYSTEMS

Elham A. Kazeeroni and Abdullah M. Al-Sadi

Department of Crop Sciences, College of Agricultural and Marine Sciences, Sultan Qaboos University, Seeb, Oman

ABSTRACT

Oasis farming system is common in some parts of the world, especially in the Arabian Peninsula and several African countries. In Oman, the farming system in the majority of farms follows a semi-oasis farming (SOF) system, which is characterized by growing multiple crops mainly for home consumption, but also for local market. This study was conducted to investigate fungal diversity using pyrosequencing approach in soils from a farm utilizing a SOF system which is cultivated with date palms, acid limes and cucumbers. Fungal diversity from this farm was compared to that from an organic farm (OR) growing cucumbers and tomatoes. Fungal diversity was found to be variable among different crops in the same farm. The observed OTUs, Chao1 richness estimates and Shannon diversity values indicated that soils from date palms and acid limes have higher fungal diversity compared to soil from cucumbers (SOF). In addition, they also indicated that the level of fungal diversity is higher in the rhizosphere of cucumbers grown in OR compared to SOF. *Ascomycota* was the most dominant phylum in most of the samples from the OR and SOF farms. Other dominant phyla are *Microsporidia*, *Chytridiomycota*, and *Basidiomycota*. The differential level of fungal diversity within the SOF could be related to the variation in the cultural practices employed for each crop.

Keywords: Oasis farming, soil-borne fungi, pesticides, organic matter, fungal communities, population dynamics

INTRODUCTION

Soil is a reservoir of thousands of fungal and bacterial species that play important roles in natural and managed agricultural soils (Abed et al., 2013; Al-Sadi et al., 2015b; Kaisermann et al., 2015; Tardy et al., 2015). Fungi are the most dominant eukaryotic species in terms of biomass in soil. Fungi play important roles as decomposers, nutrient cyclers, soil aggregators, pathogens, and mycorrhizal symbionts (Guo et al., 2015; Thomson et al., 2015; Stott and Taylor, 2016).

Changes in land use and agricultural practices have resulted in reduction in soil quality, fertility and productivity (Cherubin et al., 2015; Price et al., 2015). The productivity and health of soils rely to some extent on the processes of soil microbial communities (Guo et al., 2015; Heilmann-Clausen et al., 2015; Stott and Taylor, 2016). The high use of inorganic fertilizers and pesticides can affect soil microbial populations and result in reduction of microbial diversity or changes in microbial communities (Esmaeili Taheri et al., 2015; Filimon et al., 2015; Pose-Juan et al., 2015; Rangel et al., 2015).

Oasis farming system has been a common practice in the Arabian Peninsula over 1000s of years. The system is characterized by growing several crops, mainly date palms, around a water resource. The water resource, date palms and other crops used to provide a pleasant environment for people, where several people belonging to the same or different tribes used to live together (Kharusi and Salman, 2015). This system has been developed into a semi-oasis, conventional farming system in most growing areas in the Arabian Peninsula, especially in Oman, where farmers continued to grow multiple crops in plots which are irrigated by flowing water or underground water extracted from wells (Al-Sadi et al., 2013; Yaish and Kumar, 2015). Most farms in Oman these days have at least two or more of the following crops: date palms, acid limes, mangoes and cucumbers (Al-Sadi et al., 2013). Animal manures are commonly used in these farms, especially for date palms and citrus (Al-Sadi et al., 2011, 2015b). However, potting media, composts and several inorganic fertilizers are introduced and used for vegetable crops, mainly cucumbers (Al-Mazroui and Al-Sadi, 2015). In addition, the infection of cucumbers by fungal pathogenic fungi necessitates frequent applications of systemic and contact fungicides (Al-Sadi, 2012). On the other hand, there is a shift toward the use organic farming (OR) in some farms in the country, but the growth rate of this sector is very slow (Al-Mazroui and Al-Sadi, 2015).

Previous studies have shown that the population size and structure of soil flora and fauna can be affected by several factors, including the cultivation techniques, plant species, and the application of organic and inorganic fertilizers and pesticides (DeAngelis et al., 2015; Matsushita et al., 2015; Tardy et al., 2015; Van Geel et al., 2015; Coleman-Derr et al., 2016). However, little information is available concerning the effect of cultivation systems on fungal diversity in this part of the world. In addition, little is known about the level of fungal diversity between different crops in the same farm.

This study was conducted to investigate fungal diversity in semi-oasis and OR systems. Specific objectives are: (1) to investigate the level of fungal diversity in a semi-oasis farming (SOF) system, and (2) to characterize the extent of fungal diversity in relation to OR and in the rhizosphere of different crops. Investigations into the dynamics of microbial populations in farming systems could help provide information about the health status of soils in SOF systems, and improve the shift toward more organic-dependent commercial farming systems in the developing countries.

MATERIALS AND METHODS

Collection of Samples

The study investigated fungal diversity in a SOF system in Oman in comparison to OR. The experiment was located in one OR and one SOF. Details on the location, crops, and weather conditions are presented in Table.

Table 1: Characteristics of samples and sampling locations

Sample code	Farming system	Crop	Sampling site	Geographical location	Temperature range (°C)	Annual rain fall (mm)
OR-CU	Organic	Cucumber	Farm #1	N 23° 40.81668′, E 58° 10.95′	15.9–42.9	7 mm
OR-TO	Organic	Tomato	Farm #1	N 23° 40.81668′, E 58° 10.95′	15.9–42.9	7 mm
SOF-CU	Semi-oasis	Cucumber	Farm #2	N 23° 41.02458′, E 57° 54.29364′	13.4–43.3	21 mm
SOF-AL	Semi-oasis	Acid lime	Farm #2	N 23° 41.02458′, E 57° 54.29364′	13.4–43.3	21 mm
SOF-DP	Semi-oasis	Date palm	Farm #2	N 23° 41.02458′, E 57° 54.29364′	13.4–43.3	21 mm

Soil samples were collected during September–November, 2013, from the rhizosphere of cucumber and tomato in the OR and from cucumber, acid lime, and date palm in SOF. The rhizosphere soil in this study is considered soil which is within the 0–3 cm distance from plant roots. Cucumber and tomato have been grown in the OR for at least the last 8 years. Date palms and acid limes grown in the SOF were 9–12 years old. They have been fertilized using animal manure and have never received any fungicide treatment in the soil. Cucumbers in the SOF have been fertilized using inorganic fertilizers and animal manures. Cucumber soils from the SOF farm received several mefenoxam, hymexazol and Thiophanate-methyl treatments over the last 5 years.

Each soil sample was approximately 1kg collected from three locations in the top 5–15 cm depth near (0–3 cm) the active feeder roots of each crop. Soil samples were collected from the rhizosphere of three randomly selected crops of each crop species grown in a different plot within a farm. The soil samples were kept in sterile plastic bags and then transferred to the Plant Pathology Research Lab, SQU, Oman. Each sample was subjected to chemical and physical analysis. Soil for DNA analysis was ground with liquid nitrogen and then kept at -80°C until DNA extraction.

Soil Physicochemical Properties

Soil samples were air-dried, ground, and then passed through a 2-mm sieve to remove roots and plant debris. After that the sieved soil was stored in plastic tubes until analysis. Various physicochemical parameters of each soil sample were determined. The texture of the soil was determined by using hydrometer test (Gee and Bauder, 1986). Electrical conductivity (EC) and pH were determined by using EC and pH meters (Zhang et al., 2005). Potassium (K) contents were measured with flame photometric method (Sheerwood 450 flame photometer) while the concentration of phosphorus (P) was determined by using Inductively Coupled Plasma (Perkin Elmer, USA). Total inorganic carbon (TIC) and total organic carbon (TOC) was analyzed by using Total Organic Carbon analyzer (TOC-V, Shimadzu, Japan). Analysis of nitrogen was done by mixing 0.5 g soil sample with one tablet of Kjeltab catalyst in 10 ml of sulfuric acid, followed by heating at 420°C for 20–30 min. Then the solution was allowed to cool, followed by analysis of total nitrogen using Kjeltec Analyser (FOSS TECATOR, Sweden).

Pyrosequencing Analyses

DNA was extracted from soil samples according to the protocol of Volossiouk et al. (1995), following some modifications. Soil was ground with liquid nitrogen for 5 min and then 0.05 g of each sample was transferred to 1.5 μl eppendorf tube. Then, 125 μl of skimmed milk was added and the mixture was incubated in an oven at 65°C for 1 h. The mixture was centrifuged and the supernatant was transferred into a new eppendorf tube. After that 500 μl SDS extraction buffer (0.3 % SDS, 140 mM NaCl, 50 mM NaAc, pH 5.1) was added to the supernatant, followed by the addition of one volume of phenol: chloroform: isoamyl alcohol (25:24:1). The suspension was centrifuged and the supernatant was transferred to a new eppendorf tube. Then, 0.6 volumes of isopropanol and 10 μl of NaAc was added to the supernatant, incubated overnight at -20°C and then centrifuged. The pellet was washed with 600 μl ethanol, dried and finally suspended in TE buffer. The quality and quantity of DNA was assessed by using a Nano drop spectrophotometer (Thermo Scientific, USA).

DNA extraction was carried out from the three replicate samples collected from each crop. The three replicate DNA samples of each crop were then composted together. The primers ITS1F and ITS2aR were used for the amplification of ITS2 region. The samples were submitted to the Research and Testing Laboratory (RTL, Lubbock, TX, USA) for tag-encoded 454-pyrosequencing (Dowd et al., 2008a,b).

The obtained sequences which are less than 300 bp were excluded from further analysis and the rest were checked for high quality using RDP ver 9 (Cole et al., 2009). The low quality ends and tags were removed and were checked for chimers using UCHIME chimera detection software (Edgar et al., 2011). The resulting sequences were analyzed using a BLASTn.NET algorithm by comparison to high quality sequences from NCBI and the outputs were validated based on taxonomic distance methods (Dowd et al., 2005, 2008a,b). Out of 1980–14189 raw reads for the seven samples, 1230–11829 reads were obtained after filtering at 97% similarity threshold.

Further analysis of pyrosequencing data was carried out using the R software (R Development Core Team, 2011). A rarefaction curve plot was generated showing the number of OTUs versus the number of sequences. Richness was estimated using the Chao1 richness estimator using the formula: $S_{chao1} = S_{obs} + n_1 (n_1\text{-}1)/2 (n_2 + 1)$, where S_{obs} is the number of observed species/OTUs and ni is the number of OTUs with abundance i. Shannon Diversity was calculated using the formula $H' = -\Sigma p i \ln(pi)$, where P_i is the proportion of phylotypes, and i is the total number of all phylotypes in the sample. UniFrac and Bray–Curtis distances were calculated using the phyloseq package in R. Principal Coordinate Analyses (PCoA) were conducted and plotted from weighted and unweighted UniFrac distances and Bray–Curtis distances. Heatmap of relative abundances of the most dominant fungal genera was generated and the samples were sorted based on Bray–Curtis, weighted UniFrac and unweighted UniFrac distances.

Statistical Analysis

Tukey's Studentized range test (SAS, SAS Institute Inc., USA) was used to examine differences among soils from different cropping systems. Correlation analysis was also conducted using SAS.

RESULTS

Soil Physicochemical Properties

Soils from different crops and farming systems showed variation in their physicochemical properties (Table22). The OR soils were loamy sandy, while soils from SOF were found sandy and loamy sandy. The pH of most soils was found to be neutral to alkaline (pH = 7.7–8.4), while EC ranged from 1 to 10.54 mS. The level of inorganic carbon was found to be significantly higher in soils from SOF compared to OR, while organic carbon was found to be significantly higher in the soil from date palms compared to all other soils (*P* < 0.05). The levels of N, P, and K were variable among different farming systems and crops (Table22).

Table 2: Physicochemical analysis of soil samples

Code	Soil texture	pH	EC (mS)	%TIC	%TOC	%N	P (mg/kg)	K (mg/kg)
OR-CU	Loamy sand	8.4^{a}	0.99^{c}	1.06^{b}	1.90^{b}	0.06ˆb	3.71^{a}	7.46ˆb
OR-TO	Loamy sand	8.4^{a}	1.21^{c}	0.02^{c}	2.89^{b}	0.05ˆb	3.25^{a}	24.57ˆab
SOF-CU	Loamy sand	7.7^{b}	4.98^{b}	5.65^{a}	2.30^{b}	0.02ˆc	4.48^{a}	58.37ˆa
SOF-AL	Sandy	8.1^{ab}	2.92^{c}	5.84^{a}	2.27^{b}	0.08b	0.09^{b}	26.77ˆab
SOF-DP	Sandy	8.2^{ab}	2.20^{c}	4.98^{a}	4.93^{a}	0.12ˆa	0.51^{b}	15.36ˆb

EC, electrical conductivity; TIC, total inorganic carbon; TOC, total organic carbon; N, nitrogen; P, phosphorus; and K, potassium. Values with the same letter in the same column are not significantly different from each other at P < 0.05 (Tukey's Studentized Range test, SAS).

Relationship between Crops and Fungal Diversity

Analysis of fungal diversity in the SOF, which is growing date palm, acid lime and cucumber, showed the presence of differential diversity among the three crops. Chao1 richness estimates showed that a total of 23, 19 and 14 OTUs were detected in the rhizosphere from the three crops, respectively (Figures11 and 22). Shannon diversity estimates showed that date palm has the highest value which is 2.50, followed by acid lime (2.22) and cucumber (1.25) (Figure22).

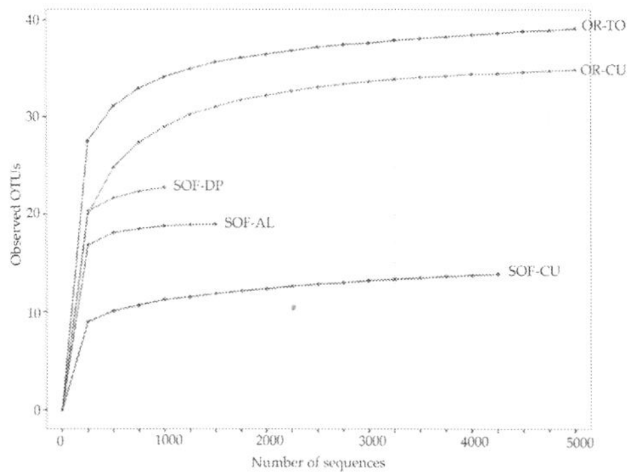

Figure 1: Rarefaction curve of observed OTUs of two soil samples obtained from tomato (OR-TO) and cucumber (OR-CU) grown in organic farm and three soil samples obtained cucumber (SOF-CU), acid lime (SOF-AL) and date palm (SOF-DP) grown in a semi-oasis farm.

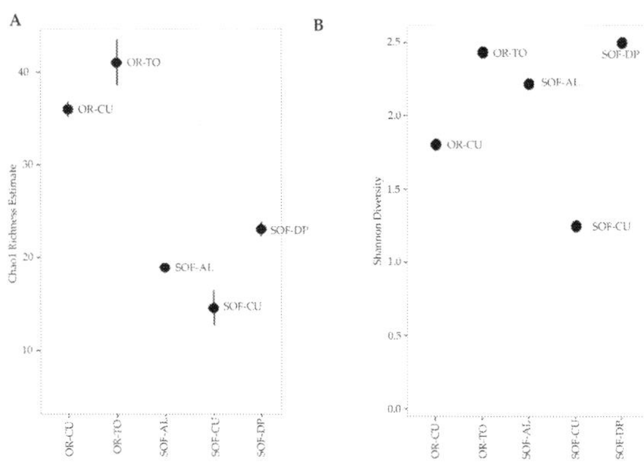

Figure 2: Chao1 richness (A) and Shannon diversity (B) within the total microbiome data of five soil samples obtained from organic farm (OR) and semi-oasis farm (SOF). Sample codes are described in Table11.

Ascomycota was the most dominant phylum, present in soil from the three crops. The other dominant phyla were *Microsporidia*, *Basidiomycota*, and *Chytridomycota* (Figure33). The classes *Dothideomycetes, Microsporidetes*, and *Sordariomycetes* were the most common in soils from the three crops (Figure33). Other common classes included *Eurotiomycetes, Leotiomycetes*, and *Chytridiomycetes*. Acid lime shared seven common classes with date palm, while cucumber shared five common classes with date palm. The number of shared classes between acid lime and cucumber were four. *Cladosporium* and *Systenostrema*were detected in soils of the three crops, while the other fungal species were detected in soil of one to two crops. *Systenostrema* had high abundance in soil from cucumber (Figure44).

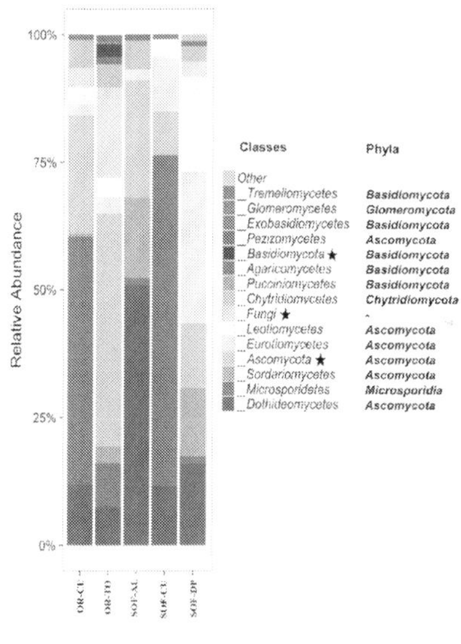

Figure 3: Class-level relative abundance of fungal communities in organic (OR) and semi-oasis farming (SOF) systems. Asterisks in the legend indicate taxonomic units that could not be resolved to the class level. The phyla of the different classes are indicated on the figure. Sample codes are described in Table11.

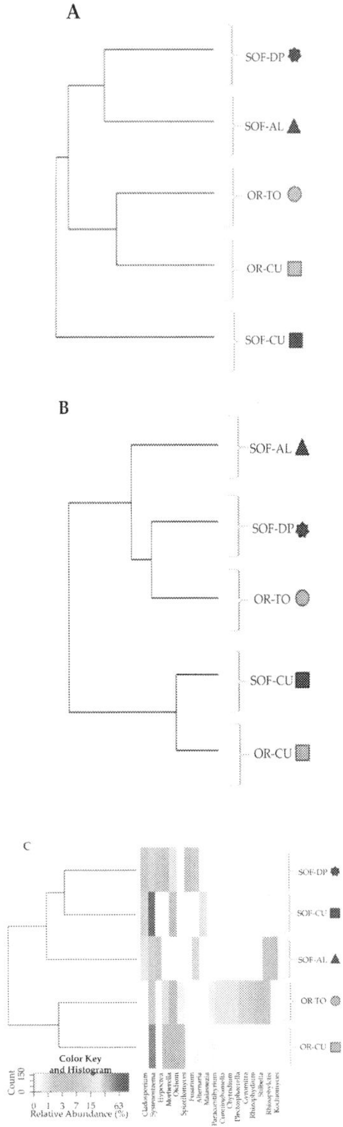

Figure 4: Heatmap of relative abundances of the 18 dominant fungal genera in the five samples. Samples are sorted based on Bray–Curtis distances (A), weighted Unifrac distances (B) and unweighted Unifrac distances (C). The colored shapes next to sample codes indicate the source of soils (green for organic and dark red for semi-oasis). The fungal genera are described in (C) while sample codes are described in Table 1.

Fungal Diversity in Organic versus Semi-Oasis Farms

A higher level of fungal diversity was observed in the rhizosphere of cucumbers grown in OR compared to SOF. The Shannon and Chao1 richness values were 1.80 and 36 for cucumber soil from OR compared to 1.25 and 14 for soil from SOF, respectively (Figure22). Tomato grown in the OR also had relatively high Shannon diversity and richness values (Figure22).

Ascomycota phylum was the most dominant phylum in most of the soil samples cultivated with cucumber in the OR and SOF as well as in the organic tomato. Other phyla included *Microsporidia*, *Chytridiomycota*, and *Basidiomycota*. Our results revealed that *Microsporidetes* was the main class in cucumber grown in OR and SOF (Figure33).

Analysis of fungal diversity in the OR and SOF soils using unweighted UniFrac distances showed separation of ORs from SOFs (Figures44 and 55). On the other hand, Weighted UniFrac separated cucumber grown in OR and SOF from other crops while Bray–Curtis also separated OR cucumbers and tomato from the others, with close relationship with SOF-DP/AL (Figures44 and 55). Further statistical analyses could not be done due to the use of one technical and one biological replicate for each sample. No correlation was found between diversity estimates and soil physicochemical properties ($P > 0.05$).

A

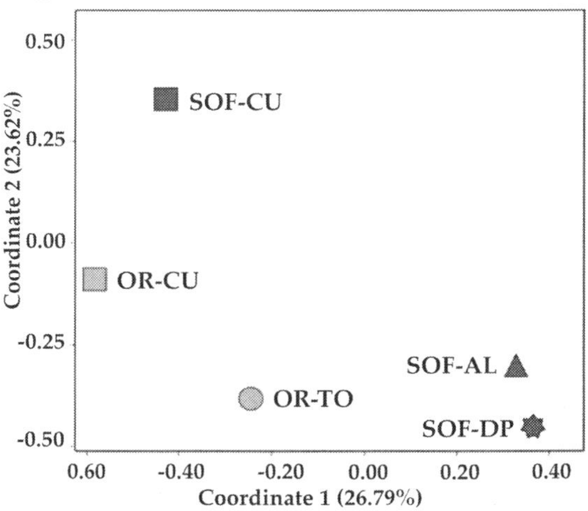

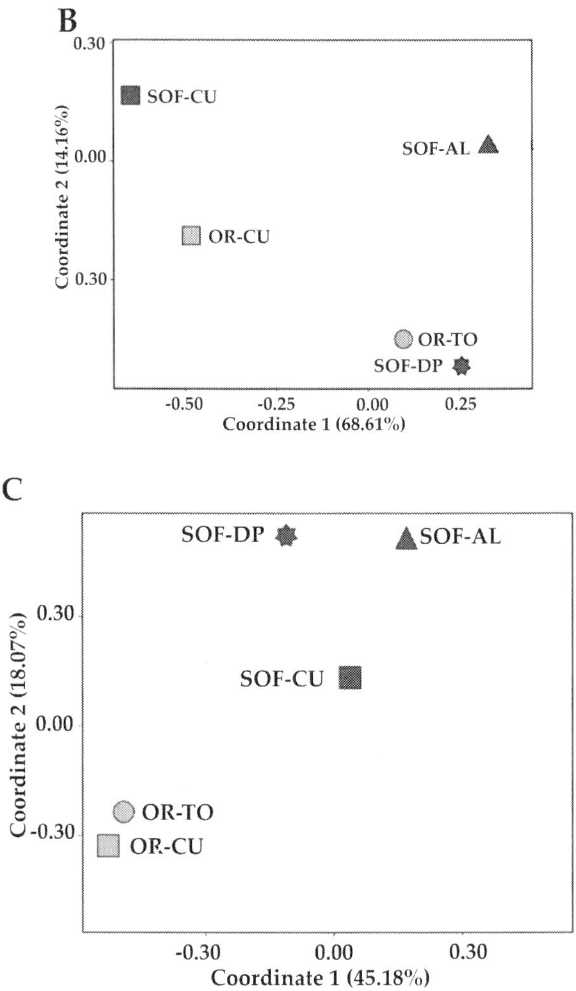

Figure 5: Principal coordinates analysis of fungal diversity in five soil samples based on Bray–Curtis distances (A), weighted Unifrac distances (B) and unweighted Unifrac distances (C). The green and dark red colors indicate organic and semi-oasis farming ...

DISCUSSION

Semi-oasis farming system in Oman has been found to harbor differential level of fungal diversity. The level of fungal diversity was higher from soils cultivated with date palms and citrus compared to soils cultivated with cucumbers. This is the first 454-pyrosequncing-based study which addressed fungal diversity in SOF. The lower level of fungal diversity in the rhizosphere of cucumber could be related to the use of fungicides in this crop. Cucumbers are heavily affected by wilt and other root diseases, which necessitate frequent applications of mefenoxam, thiophanate-methyl, hymexazol and other fungicides (Al-Sadi, 2012; Al-Sadi et al., 2012, 2015a). It is therefore likely that fungicide applications could have resulted in lowering fungal populations in the rhizosphere of cucumbers. On the other hand, fungicides' use for date palms and acid lime in Oman is minimal or does not exist at all in many farms, because of the limited fungal diseases which affect the two crops, especially date palms. In addition, tillage, which has been reported to decrease fungal diversity (Tardy et al., 2015), is frequently practiced for soils growing cucumbers, which is not the case for date palms and acid limes.

Previous studies have shown that microbial diversity increases significantly in farms which depend on organic manure as compared to chemical fertilization (Qiu et al., 2012; Zhang et al., 2012). Organic manures are known to harbor high fungal diversity (Al-Mazroui and Al-Sadi, 2015; Al-Sadi et al., 2015b). The application of animal manures instead of mineral fertilizers to date palms and acid limes could have contributed to increasing the diversity of fungi in soil from these crops. It also appears that the type of crops grown have influenced the level of fungal diversity and the interaction between these crops and fungi (Badri and Vivanco, 2009; Sugiyama et al., 2010).

The number of OTUs, Shannon diversity and Chao1 estimate revealed that the level of fungal diversity in soil from organically grown cucumber was higher compared to cucumber grown inorganically. This supports findings from previous studies about the higher fungal diversity in organic soils (Sugiyama et al., 2010). This could be related in part to the use of organic manures as well as that chemical fungicides are not used in soils in OR. In addition, organic composts which are applied frequently to these farms are usually treated with some fungal species for biocontrol purposes, which may explain the higher level of fungal diversity in these farms (Al-Sadi et al., 2015b; Kohler et al., 2015; On et al., 2015).

Unweighted UniFrac, weighted UniFrac and Bray–Curtis analyses of fungal diversity showed clustering of the different soil samples into different groups as influenced either by the faming systems and/or by plant species. Unweighted UniFrac analysis separated soils according to the farming systems, i.e., different farming systems made a qualitative difference in fungal diversity. However, weighted UniFrac showed clustering based on plant species, suggesting some dominant taxa in cucumber, which seems to be the Systenostrema (Microsporidetes). Bray–Curtis does not take into account taxonomic relatedness between OTUs, while UniFrac does. This consideration resulted in closer clustering between OR-TO with OR-CU, SOF-AL and SOF-DP, but did not affect similarity between SOF-CU and OR-CU (Figure5A5A). This shows that the OTUs that made the difference between OR-TO/CU and SOF-AL/DP detected in Bray-Curtis are somehow phylogenetically related. Also the overall variances explained by coordinate 1 and 2 is higher in weighted UniFrac (~83%), compared to unweighted (~63%) and Bray-Curtis (~50%). This may suggest that plant species and/or associated cultural practices determine the fungal community structure, although the faming systems have strong impact on the qualitative structure. In another study on *Agave* species, Coleman-Derr et al. (2016) suggested that biogeography of the host species is the major determinant of fungal microbiome.

Ascomycota phylum was the most dominant phylum in most of the samples from OR and SOF. This phylum is widespread in different soils around the world (Qiu et al., 2012; Abed et al., 2013; Al-Sadi et al., 2015b). This is mainly because *Ascomycota* contains several pigmented species which usually tolerate higher temperatures and because many species of this produce abundant spores (Abed et al., 2013). *Microsporidia*, *Chytridiomycota*, and *Basidiomycota* phyla were also common in some farm soils. Differences in the species composition between the different soils could be related to the different growing systems, cultural practices and different crops.

Physicochemical analysis of soils showed variations in soil pH, carbon and nutrients from one soil to the other. The levels of organic carbon were the same or lower in the organic soils compared to soils from semi-oasis system. Previous studies have shown that organic systems do not always increase organic matter in the soil compared to conventional systems (Li et al., 2012). The slightly lower soil pH in SOF compared to OR is in agreement with previous studies (Li et al., 2012). Soil pH and nutrients are usually affected by vegetation, soil type, carbon and nitrogen (Barak et al., 1997; Kuramae et al., 2012).

CONCLUSION

Our study showed the presence of differential levels of fungal diversity that were suggested to be associated with farming systems and plant species or cultural practices. The SOF system seems to have high level of fungal diversity, especially under crops which are grown in a semi-organic way. This is expected to be reflected on soil health and quality especially if the fungal communities are of the plant growth promoting or disease suppressive species. The trend toward production of cucumber in these systems appears to have impact on the fungal diversity levels, possibly due to the heavy use of fungicides. The current shift toward the use of pesticides in SOF systems should take into consideration the negative side of chemicals on diversity of the beneficial fungi.

AUTHOR CONTRIBUTIONS

AA-S planned for research, analyzed data, wrote paper. EK did the experiments, analyzed data and helped in writing paper.

CONFLICT OF INTEREST STATEMENT

The authors declare that the research was conducted in the absence of any commercial or financial relationships that could be construed as a potential conflict of interest.

ACKNOWLEDGMENTS

Authors would like to acknowledge Mr. Issa Al-Mahmooli and Mr. Mahmoud Al-Ansari for their help in the collection of soil samples.

REFERENCES

1. Abed R. M. M., Al-Sadi A. M., Al-Shihi M. A., Al-Hinai S., Robinson M. D. (2013). Diversity of free-living and lichenized fungal communities in biological desert crusts of the Sultanate of Oman and their role in improving soil properties. *Soil Biol. Biochem.*

57 695–705. 10.1016/j.soilbio.2012.07.023

2. Al-Mazroui S. S., Al-Sadi A. M. (2015). Highly variable fungal diversity and the occurrence of potentially plant pathogenic fungi in potting media, organic fertilizers and composts originating from 14 countries. *J. Plant Pathol.* 97 529–534. 10.4454/JPP.V97I3.033
3. Al-Sadi A. M. (2012). Efficacy of mefenoxam is affected by a lag period between application and inactivation of *Pythium* species. *Phytopathol. Mediterran.* 51 292–297.
4. Al-Sadi A. M., Al-Ghaithi A. G., Al-Balushi Z. M., Al-Jabri A. H. (2012). Analysis of diversity in *Pythium aphanidermatum* populations from a single greenhouse reveals phenotypic and genotypic changes over 2006 to 2011. *Plant Dis.* 96 852–858. 10.1094/PDIS-07-11-0624
5. Al-Sadi A. M., Al-Masoodi R. S., Al-Ismaili M., Al-Mahmooli I. H. (2015a). Population structure and development of resistance to hymexazol among *Fusarium solani* populations from date palm, Citrus and cucumber. *J. Phytopathol.* 163 947–955. 10.1111/jph.12397
6. Al-Sadi A. M., Al-Mazroui S. S., Phillips A. (2015b). Evaluation of culture-based techniques and 454 pyrosequencing for the analysis of fungal diversity in potting media and organic fertilizers. *J. Appl. Microbiol.* 119 500–509. 10.1111/jam.12854
7. Al-Sadi A. M., Al-Said F. A., Al-Jabri A. H., Al-Mahmooli I. H., Al-Hinai A. H., de Cock A. W. A. M. (2011). Occurrence and characterization of fungi and oomycetes transmitted via potting mixtures and organic manures. *Crop Prot.* 30 38–44. 10.1016/j.cropro.2010.09.015
8. Al-Sadi A. M., Al-Wehaibi A. N., Al-Shariqi R. M., Al-Hammadi M. S., Al-Hosni I. A., Al-Mahmooli I. H., et al. (2013). Population genetic analysis reveals diversity in *Lasiodiplodia* species infecting date palm, Citrus, and mango in Oman and the UAE. *Plant Dis.* 97 1363–1369. 10.1094/PDIS-03-13-0245-RE
9. Badri D. V., Vivanco J. M. (2009). Regulation and function of root exudates. *Plant Cell Environ.* 32666–681. 10.1111/j.1365-3040.2009.01926.x
10. Barak P., Jobe B. O., Krueger A. R., Peterson L. A., Laird D. A. (1997).

Effects of long-term soil acidification due to nitrogen fertilizer inputs in Wisconsin. *Plant Soil* 197 61–69. 10.1023/a:1004297607070

11. Cherubin M. R., Franco A. L. C., Cerri C. E. P., Oliveira D. M. S., Davies C. A., Cerri C. C. (2015). Sugarcane expansion in Brazilian tropical soils-Effects of land use change on soil chemical attributes. *Agricult. Ecosyst. Environ.* 211 173–184. 10.1016/j.agee.2015.06.006
12. Cole J. R., Wang Q., Cardenas E., Fish J., Chai B., Farris R. J., et al. (2009). The Ribosomal Database Project: improved alignments and new tools for rRNA analysis. *Nucleic Acids Res.* 37D141–D145. 10.1093/nar/gkn879
13. Coleman-Derr D., Desgarennes D., Fonseca-Garcia C., Gross S., Clingenpeel S., Woyke T., et al. (2016). Plant compartment and biogeography affect microbiome composition in cultivated and native Agave species. *New Phytol.* 209 798–811. 10.1111/nph.13697
14. DeAngelis K. M., Pold G., Topçuoglu B. D., van Diepen L. T. A., Varney R. M., Blanchard J. L., et al. (2015). Long-term forest soil warming alters microbial communities in temperate forest soils. *Front. Microbiol.* 6:104 10.3389/fmicb.2015.00104
15. Dowd S. E., Callaway T. R., Wolcott R. D., Sun Y., McKeehan T., Hagevoort R. G., et al. (2008a). Evaluation of the bacterial diversity in the feces of cattle using 16S rDNA bacterial tag-encoded FLX amplicon pyrosequencing (bTEFAP). *BMC Microbiol.* 8:125 10.1186/1471-2180-8-125
16. Dowd S. E., Sun Y., Wolcott R. D., Domingo A., Carroll J. A. (2008b). Bacterial tag encoded FLX amplicon pyrosequencing (bTEFAP) for microbiome studies:bacterial diversity in the ileum of newly weaned *Salmonella*-infected pigs. *Foodborne Pathog. Dis.* 5 459–472. 10.1089/fpd.2008.0107
17. Dowd S. E., Zaragoza J., Rodriguez J. R., Oliver M. J., Payton P. R. (2005). Windows.NET network distributed basic local alignment search toolkit (W.ND-BLAST). *BMC Bioinformat.* 6:9310.1186/1471-2105-6-93
18. Edgar R. C., Haas B. J., Clemente J. C., Quince C., Knight R. (2011). UCHIME improves sensitivity and speed of chimera detection. *Oxford J. Bioinformat.* 27 2194–2200. 10.1093/bioinformatics/

btr381

19. Esmaeili Taheri A., Hamel C., Gan Y. (2015). Pyrosequencing reveals the impact of foliar fungicide application to chickpea on root fungal communities of durum wheat in subsequent year. *Fungal Ecol.* 15 73–81. 10.1016/j.funeco.2015.03.005
20. Filimon M. N., Voia S. O., Popescu R., Dumitrescu G., Ciochina L. P., Mituletu M., et al. (2015). The effect of some insecticides on soil microorganisms based on enzymatic and bacteriological analyses. *Roman. Biotechnolog. Lett.* 20 10439–10447.
21. Gee G. W., Bauder J. W. (1986). "Particle size analysis," in *Methods of Soil Analysis, Part-I, Physical and Mineralogical Methods*, ed. Klute A., editor. (Madison, WI: American Society of Agronomy;), 383–411.
22. Guo X., Petermann J. S., Schittko C., Wurst S. (2015). Independent role of belowground organisms and plant cultivar diversity in legume-grass communities. *Appl. Soil Ecol.* 95 1–8. 10.1016/j.apsoil.2015.05.010
23. Heilmann-Clausen J., Barron E. S., Boddy L., Dahlberg A., Griffith G. W., Nordén J., et al. (2015). A fungal perspective on conservation biology. *Conservation Biol.* 29 61–68. 10.1111/cobi.12388
24. Kaisermann A., Maron P. A., Beaumelle L., Lata J. C. (2015). Fungal communities are more sensitive indicators to non-extreme soil moisture variations than bacterial communities. *Appl. Soil Ecol.* 86 158–164. 10.1016/j.apsoil.2014.10.009
25. Kharusi N. S., Salman A. (2015). In search of water: hydrological terms in Oman's toponyms.*Names* 63 16–29. 10.1179/0027773814Z.00000000094
26. Kohler J., Caravaca F., Azcón R., Díaz G., Roldán A. (2015). The combination of compost addition and arbuscular mycorrhizal inoculation produced positive and synergistic effects on the phytomanagement of a semiarid mine tailing. *Sci. Total Environ.* 514 42–48. 10.1016/j.scitotenv.2015.01.085
27. Kuramae E. E., Yergeau E., Wong L. C., Pijl A. S., Van Veen J. A., Kowalchuk G. A. (2012). Soil characteristics more strongly influence soil bacterial communities than land-use type. *FEMS Microbiol. Ecol.* 79 12–24. 10.1111/j.1574-6941.2011.01192.x

28. Li R., Khafipour E., Krause D. O., Entz M. H., de Kievit T. R., Fernando W. G. D. (2012). Pyrosequencing reveals the influence of organic and conventional farming systems on bacterial communities. *PLoS ONE* 7:e51897 10.1371/journal.pone.0051897

29. Matsushita Y., Bao Z., Kurose D., Okada H., Takemoto S., Sawada A., et al. (2015). Community structure, diversity, and species dominance of bacteria, fungi, and nematodes from naturally and conventionally farmed soil: a case study on Japanese apple orchards. *Organ. Agricult.* 5 11–28. 10.1007/s13165-015-0096-4

30. On A., Wong F., Ko Q., Tweddell R. J., Antoun H., Avis T. J. (2015). Antifungal effects of compost tea microorganisms on tomato pathogens. *Biol. Cont.* 80 63–69. 10.1016/j.biocontrol.2014.09.017

31. Pose-Juan E., Sánchez-Martín M. J., Herrero-Hernández E., Rodríguez-Cruz M. S. (2015). Application of mesotrione at different doses in an amended soil: dissipation and effect on the soil microbial biomass and activity. *Sci. Total Environ.* 536 31–38. 10.1016/j.scitotenv.2015.07.039

32. Price G. W., Astatkie T., Gillis J. D., Liu K. (2015). Long-term influences on nitrogen dynamics and pH in an acidic sandy soil after single and multi-year applications of alkaline treated biosolids. *Agricult. Ecosys. Environ.* 208 1–11. 10.1016/j.agee.2015.04.010

33. Qiu M., Zhang R., Xue C., Zhang S., Li S., Zhang N., et al. (2012). Application of bio-organic fertilizer can control Fusarium wilt of cucumber plants by regulating microbial community of rhizosphere soil. *Biol. Fertil. Soils* 48 807–816. 10.1002/jsfa.5653

34. R Development Core Team (2011). *R: A Language and Environment for Statistical Computing.*Vienna: R Foundation for Statistical Computing.

35. Rangel D. E. N., Alder-Rangel A., Dadachova E., Finlay R. D., Kupiec M., Dijksterhuis J., et al. (2015). Fungal stress biology: a preface to the fungal stress responses special edition. *Curr. Genet.*61 231–238. 10.1007/s00294-015-0500-3

36. Stott M. B., Taylor M. W. (2016). Microbial ecology research in New Zealand. *N. Z. J. Ecol.* 40 12–28.

37. Sugiyama A., Vivanco J. M., Jayanty S. S., Manter D. K. (2010). Pyrosequencing assessment of soil microbial communities in

organic and conventional potato farms. *Plant Dis.* 94 1329–1335. 10.1094/PDIS-02-10-0090

38. Tardy V., Spor A., Mathieu O., Lévèque J., Terrat S., Plassart P., et al. (2015). Shifts in microbial diversity through land use intensity as drivers of carbon mineralization in soil. *Soil Biol. Biochemist.*90 204–213. 10.1016/j.soilbio.2015.08.010
39. Thomson B. C., Tisserant E., Plassart P., Uroz S., Griffiths R. I., Hannula S. E., et al. (2015). Soil conditions and land use intensification effects on soil microbial communities across a range of European field sites. *Soil Biol. Biochemist.* 88 403–413. 10.1016/j.soilbio.2015.06.012
40. Van Geel M., Ceustermans A., Van Hemelrijck W., Lievens B., Honnay O. (2015). Decrease in diversity and changes in community composition of arbuscular mycorrhizal fungi in roots of apple trees with increasing orchard management intensity across a regional scale. *Mol. Ecol.* 24 941–952. 10.1111/mec.13079
41. Volossiouk T., Robb E. J., Nazar R. N. (1995). Direct DNA extraction for PCR-mediated assays of soil organisms. *Appl. Environ. Microbiol.* 61 3972–3976.
42. Yaish M. W., Kumar P. P. (2015). Salt tolerance research in date palm tree (Phoenix dactylifera L.), past, present, and future perspectives. *Front. Plant Sci.* 6:348 10.3389/fpls.2015.00348
43. Zhang H., Schroder L. L., Pittman J. J., Wang J. J., Payton M. E. (2005). Soil salinity using saturated paste and 1;1 soil and water extracts. *Soil Sci. Soc. Am. J.* 69 1146–1151. 10.2136/sssaj2004.0267
44. Zhang Q. C., Shamsi I. H., Xu D. T., Wang G. H., Lin X. Y., Jilani G., et al. (2012). Chemical fertilizer and organic manure inputs in soil exhibit a vice versa pattern of microbial community structure. *Appl. Soil Ecol.* 57 1–8. 10.1016/j.apsoil.2012.02.012

CHAPTER 8

DEVELOPMENT OF A PYROSEQUENCING ASSAY FOR THE TYPING OF ALPHAHERPESVIRUSES

G. Fusco [a], M.G. Amoroso [a], N. Gesualdi Montesano [b], and M. Viscardi [a]

[a]Department of Animal Health, Experimental Zooprophylactic Institute of Southern Italy, Via Salute, 2, 80055 Portici (NA), Italy

[b]Qiagen Srl, Via Grosio 10/8, 20151 Milano, Italy

ABSTRACT

Identification of herpesvirus in biological material is usually carried out by real-time PCR. With the aim to classify the strain of virus identified, real-time PCR must be often supported by time-consuming capillary electrophoresis sequencing analysis. Here we provide a protocol for the rapid and reliable identification of 5 closely related herpesviruses by PyroMark Q24 sequencing system. PyroMark performs DNA sequencing analysis using pyrosequencing, a technology based on the detection of released pyrophosphate during DNA elongation [1]. PyroMark is designed to detect changes in specified variable positions of the DNA. It can efficiently detect single nucleotide differences in sequences [2]. In the present paper we describe a protocol to pyrosequence a small polymorphic segment of the US8 gene. On the basis of the differences identified in the nucleotide sequence we could readily classify the herpesvirus as *Bovine herpesvirus 1.1, Bovine herpesvirus 1.2, Bovine herpesvirus 5, Bubaline herpesvirus 1* or *Caprine herpesvirus.* The protocol set up offers several advantages with respect to the techniques commonly used:

- it requires less than one working day to be carried;
- it gives the possibility to analyze, at reasonable costs, up to 24 samples at a time; and
- it allows to detect with great reliability and specificity strongly genetically correlated organisms like the herpesviruses named above. The procedure can be easily applied to other families of viruses, with opportune modifications.

Keywords: Pyrosequencing, Alphaherpesviruses, Polymorphic site, PyroMark, Biotinylated primer, US8 gene

METHOD DETAILS

In short the typing of the alphaherpesviruses by pyrosequencing is made as follows: (A) extraction of DNA from the samples; (B) PCR amplification with two amplification primers (forward and reverse) that are specific to the sequence of interest, one of which (in our case the reverse primer) is biotinylated; (C) capture of biotin-labeled single-stranded PCR products (ssDNA) and separation from non-biotin-labeled strands; (D) incubation of the ssDNA with the sequencing primer; (E) sequencing of the variable region by pyrosequencing; (F) analysis of the pyrograms obtained and classification of the strain of herpesvirus.

EXPERIMENTAL DESIGN

Pyrosequencing assay involves 3 primers: forward, reverse and sequencing. To design the primers we first performed a multiple alignment (ClustalW alignment software) of the US8 gene of the 5 herpesviruses (Fig. 1) and, on the basis of the alignment, we built up the pyrosequencing assay (Fig. 2). As it is possible to see from Fig. 2, forward and reverse primers has been designed to amplify an amplicon (302 bp) common to all the herpesviruses in study and containing a short sequence variable among the strains (polymorphic site). The sequencing primer was designed to be located just upstream the polymorphic region in order to analyze downstream nucleotide segment. All the primers were designed with the PyroMark Assay Design Software 2.0 (Qiagen) setting the assay type as "sequence analysis (SQA)". After primers design, with the help of the PyroMark Q24 2.0.6 software we built up the pyrosequencing running protocol (PRP), establishing the set up of the plate and the PyroMark SQA assay. The software, on the basis of the information filled in (including the number of samples to analyze), supplies the total volume of reagents to add to the cartridge (see Fig. 3) and the order of cyclic nucleotides dispensation (8x[ACGT]). All the assay information is saved on an electronic key and transferred to the PyroMArk Q24 station where the assay is run.

```
                            .|....|....|....|....|
                            360       370       380
Bovine herpesvirus 1.1      GA GGCCGCAGCC GCCGACGCTC
Bovine herpesvirus 1.2      GA GGCCTCAGCC GCCGACGCTC
Bubaline herpesvirus 1      GA GGCCGCGGCC GCCGACGCCC
Bovine herpesvirus 5        G- -----CGGCC GCCGACGCCC
Caprine herpesvirus 1       GA GGCCGCGGGC GCCGACGGCG
```

Figure 1: ClustalW alignment of the polymorphic site of the 5Herpesvirus US8 sequences. In grey the nucleotide differences with respect to *Bovine herpesvirus 1.1*. The alignment was carried out with the Bioedit sequence alignment editor software.

```
         F                                                                                S
    TGCTGCAGTA CAACGGCCAC GT                                                       ACC GACCACACGC
    |||||||||| |||||||||| ||                                                       ||| ||||||||||
264 TGCTGCAGTA CAACGGCCAC GTGGAAGCTT GGGACTACAG CCTAGTCGTT ACTTCGGACC GTTTGGTGCG CGCGGTCACC GACCACACGC

    GCCCC
    |||||
354 GCCCCGAGGC CGCAGCCGCC GACGCTCCCG AGCCAGGCCC ACCGCTCACC AGCGAGCCGG CGGGCGCGCC CACCAGGCCC GCGCCCTGGC

444 TTGTGGTGCT GGTGGGCGCG CTTGGACTCG CGGGACTGGT GGGCATCGCR GCCCTCGCCG TTCGGTGTG CGCGCGCCGC GCAAGCCAGA

534 AGCGCACC TACGACATCC TCAACCCCTT CGGGC
                    ||| |||||||||| |||||
                    AGG AGTTGGGGAA GCCCG

                          R
```

Figure 2: Schematic representation of the pyrosequencing assay with forward (F), reverse (R) and sequencing (S) primers. The sequence of DNA refers to a region (302 pb) of the gene US8 of the *Bovine herpesvirus*1.1.

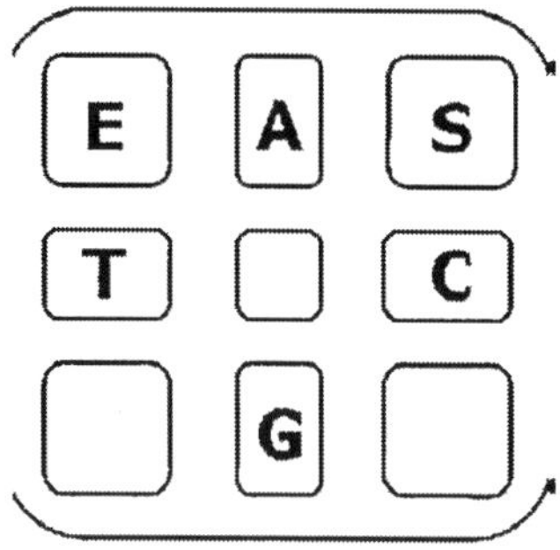

Figure 3: PyroMark Q24 cartridge scheme of loading. The amount of reagents to add per sample is the following: enzyme mix (E): 54 µl; substrate mix (S) 54 µl, nucleotide A (A) 52 µl, nucleotide C (C) 52 µl, nucleotide G (G) 52 µl, nucleotide T (T) 52µl.

Step 1: nucleic acid extraction

Materials

Viral strains used to validate the method

Reference strains of *BoHV-1.1*, *BoHV-1.2*, *BoHV-5* and *CpHV-1*were supplied by Dr. M. Beer from FLI (Friedrich Loeffler Institut, Germany). *Bubaline herpesvirus* 1 (strain Metzler) was kindly provided by Prof. U. Pagnini (University of Naples Federico II, Italy).

- 3–7 mm stainless steel beads
- Disposable scalpel
- Sterile pincers and scissors
- QIAamp viral RNA mini kit (Qiagen)
- Technical scale
- Tissue Lyser (Qiagen)

The protocol is described for animal tissues but can be easily applied to samples like plasma, urine, cell culture, swabs and blood. It is possible to work with lower/higher amount of tissue paying attention to maintain the ratio 1:10 tissue/PBS. If processing fiber-rich tissues, complete disruption and homogenization may sometimes not be possible. However, small amounts of debris have no effect on subsequent nucleic acid extraction.

Procedure

- Cut the tissue of each sample under analysis in small pieces with the help of the scalpel, the pincers and the scissors.
- Weight 150 mg of tissue and place it in a 2 ml microcentrifuge tube.
- Add 1 stainless steel bead (3–7 mm mean diameter) and 1.5 ml of PBS.
- Place the tubes in the TissueLyser and omogenize for 5 min at 30 Hz.
- The duration of disruption and homogenization depends however on the tissue being processed and can be extended until no tissue debris is visible.
- Centrifuge the samples 15 min/4000 × *g*/4 °C.
- Transfer the supernatant in a new 2 ml microcentrifuge tube.
- Process 140 µl of the supernatant with QIAamp viral RNA mini kit following handbook instructions.
- Elute nucleic acids in 50 µl of QIAamp elution buffer.
- Amplify immediately extracted DNA or store it at −20 °C until use.

Step 2: PCR amplification

Materials

- Forward specific primer: 5′-TGCTGCAGTACAACGGCCACGT-3′;
- Reverse biotinylated specific primer: 5′-GCCCGAAGGGGTTGAGGA-3′.
- PyroMark PCR master mix (Qiagen)
- Qsolution
- PCR amplification system
- Capillary electrophoresis system

Procedure

- Prepare, in a 1.5 ml microcentrifuge tube, the PCR reaction mix including per sample: 12.5 µl PyroMark PCR master mix, 1.25 µl of each primer (0.2 µM), 5 µl Qsolution.
- Distribute the mix (20 µl/sample) in 0.2 ml PCR tubes
- Add 5 µl of sample (extracted DNA) to the relative tube.
- Perform PCR amplification with the following thermal profile: a first activation step of 15 min at 95 °C, 45 cycles of: 30 s at 94 °C, 30 s at 60 °C, and 30 s at 72 °C and a final elongation step of 10 min at 72 °C.
- Check PCR product by capillary electrophoresis. Conventional agarose gel electrophoresis can be carried out if capillary electrophoresis system is not available. *Note*: if non specific amplicons also result from the PCR it is necessary to perform further optimization of the PCR conditions.

Step 3: preparation of the samples and pyrosequencing

Materials

- PyroMark Q24 Vacuum workstation (Qiagen)
- Plate mixer for immobilization to beads
- Heating block capable of attaining 80 °C
- PyroMark Q24 plate (Qiagen)
- PyroMark Plate Holder (Qiagen)
- PyroMark Q24Cartridge (Qiagen)
- Streptavidin Sepharose High Performance (GE Healthcare)
- Sequencing specific primer:
  5′-ACCGACCACACGCGCCCC-3′
- PyroMark Binding buffer
- PyroMark Denaturation solution
- PyroMark Wash buffer concentrate
- PyroMark Annealing buffer
- PyroMark Gold Q24 reagents (containing enzymes, substrates and nucleotides for pyrosequencing)

Procedure

- Prepare, in a 2 ml microcentrifuge tube, the immobilization solution by mixing (per sample): 2 μl steptavidin-coated Sepharose beads together with 40 μl Binding buffer and 18 μl high-purity water.
- Transfer biotinylated amplified samples (20 μl) to the wells of a PCR plate.
- Add 60 μl of immobilization solution to the samples in the PCR plate.
- Seal the plate and capture PCR products on the steptavidin-coated beads by shaking the plate on a plate mixer for 10 min at 1400 rpm.
- In the meanwhile prepare the vacuum workstation by filling the 5 troughs with the following solutions:
  - 50 ml ethanol 70%
  - 40 ml Denaturation solution
  - 50 ml 1x Washing buffer
  - 50 ml high-purity water
  - 70 ml high-purity water
- Immediately after immobilization perform separation of biotinylated PCR strands from non-biotinylated strands in the vacuum workstation as indicated in the PyroMark user manual. *Note*: Beads sediment quickly, so be careful to perform this step right away after the capture.
- At the end of the procedure samples will be realized into a PyroMArk Q24 (PMQ) plate which contains 25 μl (per well) of the sequencing primer (opportunely diluted to 0.3 μM in the Annealing buffer).
- Accommodate the PMQ plate into the PyroMark Plate Holder
- Perform annealing of the primer to the samples by heating the PyroMark plate at 80 °C for 2 min and cooling it for 5 min at room temperature.
- Prepare the Gold Q24 reagents (enzyme and substrate mixtures) as indicated by the manufacturer.

- Select the PRP in the PyroMark Q24 2.0.6 software and set the number of samples and the scheme of the PMQ plate.
- Transfer, with the help of the electronic key, the assay information on the PyroMark Q24 station.
- Pipet the PRP established amount of reagents into the cartridge compartments according to Fig. 3.
- Place the PMQ plate and the cartridge in the instrument and start the run.
- Analyze obtained pyrograms (Fig. 4) with the PMQ24 sequencing software (Qiagen).

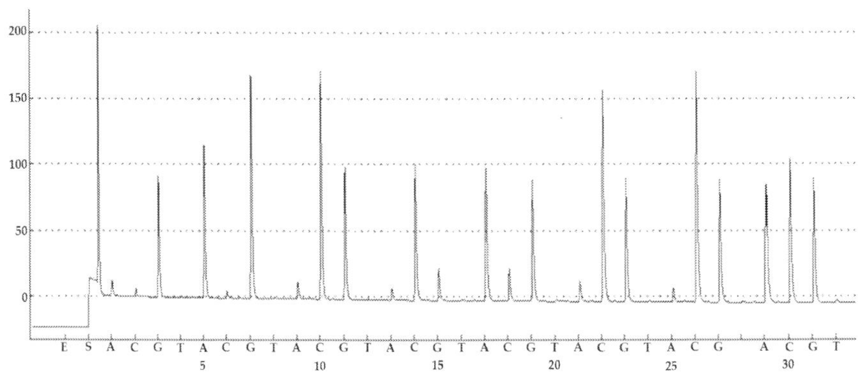

Figure 4: SQA pyrogram report run. Example of the pyrogram resulting from *Bubaline Herpesvirus 1*. Sequence read: GAGGCCGCAG CCGCCGACG.

- Classify the herpesvirus comparing the sequence obtained to those expected for each strain. *Note:* The lists of materials include only non standard items. General materials like microcentrifuge tubes, or PBS are assumed to be available. For PCR and pyrosequencing steps it is recommended to use only PCR clean disposable items.

ADDITIONAL INFORMATION

Background

The PyroMark Q24 system uses pyrosequencing technology for real-time, sequence-based detection of sequence variants, being able to analyze segments of DNA up to 100 nucleotides in length [3], [4].Pyrosequencing technique has already been successfully employed for the typing of herpes simplex virus types 1 and 2 [3], as well as human papillomavirus [5]. During Pyrosequencing reaction the sequencing primer is extended to provide the sequence of the template DNA. Nucleotides are added one at a time in a predefined order. The first nucleotide is added and if it is complementary to the base in the template strand it will be incorporated into the elongating DNA strand by DNA polymerase. The incorporation is accompanied by release of pyrophosphate (PPi). PPi is converted to ATP by ATP sulfurylase, in the presence of adenosine 5′ phosphosulfate. This reaction catalyses the conversion of luciferin to oxyluciferin by luciferase, with the emission of visible light that is recorded by a LCC camera. The complementary sequence is built up and each nucleotide incorporation generates light seen as a peak in the Pyrogram. The sequence is determined, at the end of the run, from the peaks in the Pyrogram.

REFERENCES

1. Royo J.L., Hidalgo M., Ruiz A. Pyrosequencing protocol using a universal biotinylated primer for mutation detection and SNP genotyping. Nat. Protoc. 2007;2:1734–1739.
2. Tewari D., Del Piero F., Cieply S., Feria W., Acland H. Equine herpesvirus1 (EHV-1) nucleotide polymorphism determination using formalin fixed tissues in EHV-1 induced abortions and myelopathies with real-time PCR and pyrosequencing. J. Virol. Methods. 2013;193:371–373.
3. Adelson M.E., Feola M., Jason T., Tilton R.C., Mordechai E. Simultaneous detection of herpes simplex virus types 1 and 2 by real-time PCR and pyrosequencing. J. Clin. Virol. 2005;33:25–34.
4. Gharizadeh B., Nordstrom T., Ahmadian A., Ronaghi M., Nyren P. Long-read pyrosequencing using pure 2′-deoxyadenosine-5′-O′-(1-thiotriphosphate) sp-isomer. Anal. Biochem. 2002;301:82–90.
5. Gharizadeh B., Kalantari M., Garcia C.A., Johansson B., Nyrén P. Typing of human papillomavirus by pyrosequencing. Lab. Invest. 2001;815:673–679.

CHAPTER 9

EXPLORING MICROBIAL DIVERSITY USING 16S RRNA HIGH-THROUGHPUT METHODS

Fabrice Armougom and Didier Raoult

URMITE - UMR CNRS 6236, IRD 3R198, Université de la Méditerranée, Faculté de Médecine, 27 Boulevard Jean Moulin, 13005 Marseille, France

ABSTRACT

As a result of advancements in high-throughput technology, the sequencing of the pioneering 16S rRNA gene marker is gradually shedding light on the taxonomic characterization of the spectacular microbial diversity that inhabits the earth. 16S rRNA-based investigations of microbial environmental niches are currently conducted using several technologies, including large-scale clonal Sanger sequencing, oligonucleotide microarrays, and, particularly, 454 pyrosequencing that targets specific regions or is linked to barcoding strategies. Interestingly, the short read length produced by next-generation sequencing technology has led to new computational efforts in the taxonomic sequence assignment process. From a medical perspective, the characterization of the microbial composition of the skin surface, oral cavity, and gut in both healthy and diseased people enables a comparison of microbial community profiles and also contributes to the understanding of the potential impact of a particular microbial community.

INTRODUCTION

Until recently, the vast majority of global microbial diversity was inaccessible or largely underestimated by culture dependent methods, since the cultivated fraction of the 4- 6x1031 prokaryotic genomes moving around the biosphere (Whitman et al, 1998) is currently estimated to be 1% (Giovannoni and Stingl, 2005). However, the development of culture-independent methods and the commercialization of next-generation sequencing technology (Mardis, 2008, Rothberg and Leamon, 2008) have yielded powerful new tools in terms of time savings, cost effectiveness, and data production capability. These new tools allow for the gradual characterization of the unseen majority of environmental microbial communities. Microbial diversity has recently been explored in a great variety of environments, including soil (Roesch et al, 2007, Yergeau et al, 2008), sea (Huber et al, 2007, Sogin et al, 2006), air (Wilson et al, 2002), and the human body, including from a medical perspective, the gastrointestinal tract (Andersson et al, 2008, Ley et al, 2006), oral cavity (Jenkinson and Lamont, 2005), vaginal tract (Zhou et al, 2004), and skin surface (Fierer et al, 2008). These microbial communities have been characterized in terms of community structure, composition, metabolic function, and ecological roles. Investigations of environmental microbial diversity have employed the 16S rRNA (16S) gene marker, which offers phylogenetic taxonomic classification without requiring isolation and cultivation. Although the use of the 16S phylogenetic marker is often criticized, due to its heterogeneity among operons of the same genome (Acinas et al, 2004) or its lack of resolution at the species level (Pontes et al, 2007), it is still considered as a 'gold standard' for bacterial identification. The use of next-generation sequencing technology has increased the size of 16S sequence databases at an impressive speed (Tringe and Hugenholtz, 2008).

Supported by new high-throughput methods (454 pyrosequencing, PhyloChip microarrays) and strategies (barcoding); the surveys of 16S gene in the human microbiota attempt to provide a comprehensive picture of the community differences between healthy and diseased states. In this review, we focus on the 16S-gene-based characterization of microbial communities using clonal Sanger sequencing, phylogenetic oligonucleotide microarrays, and 454 pyrosequencing strategies as applied to medical research. Propelled by the launch of the Human Microbiome Project, 16S high-throughput methods show tremendous potential for identifying uncultivated or rare pathogenic agents, finding shifts in the bacterial community associated with disease states (Ley et al, 2006), understanding how microbiota are affected by environmental factors of the human host (diet, lifestyle, sex, age) (Fierer et al, 2008), and differentiating between a core human microbial community and inter-individual variability (Gao et al, 2007, Turnbaugh et al, 2008). These advances will contribute to a more comprehensive picture of both healthy and diseased states and will lead to the use of more appropriate medical treatments, such as targeted antibiotic therapy rather than the use of broad-range antibiotics.

BACTERIAL TAXONOMIC CLASSIFICATION

The 16S rRNA gene: a phylogenetic marker

In the mid-1980s, major enhancements in bacterial typing and characterization of phylogenetic relationships were achieved, using new molecular approaches based on full length sequencing of ribosomal genes. Pioneering work by Woese and colleagues (Woese, 1987) described bacterial rRNA genes as 'molecular clocks', due to their uncommon features such as universality, activity in cellular functions, and extremely conserved structure and nucleotide sequence. The three types of rRNA in prokaryotic ribosomes are classified as 23S, 16S, and 5S, according to their sedimentation rates, and have sequence lengths of about 3300, 1550, and 120 nucleotides, respectively (Rossello-Mora and Amann,2001). Initially, microbial diversity studies involved sequencing the 5S rRNA gene obtained from environmental samples (Lane et al, 1985, Stahl et al, 1985). However, the relatively short sequence length of the 5S gene contains few phylogenetically informative sites, which limits its usefulness for taxonomic classification purposes. In addition, although the information content of the 23S rRNA gene is larger than that of the 16S gene, it is the 16S gene that has become a standard in bacterial taxonomic classification because it is more easily and rapidly sequenced (Spiegelman et al, 2005). It is widely accepted that a compelling classification of prokaryotes should be based on a 'polyphasic approach' that includes genomic, phenotypic, and phylogenetic information (Vandamme et al, 1996). However, most bacterial diversity surveys exclusively target the 16S gene in a single step phylogenetic approach (Pace, 1997).

The 1550 base pairs of the 16S gene are a structural part of the 30S ribosomal small subunit (SSU) and consist of eight highly conserved regions (U-U8) and nine variable regions across the bacterial domain (Jonasson et al, 2002). As no lateral gene transfer seems to occur between 16S genes (Olsen et al, 1986) and as their structure contains both highly conserved and variable regions with different evolution rates, the relationships between 16S genes reflect evolutionary relationships between organisms. A comparison of 16S gene sequence similarities is usually used as the 'gold standard' for taxonomic identification at the species level. Although thresholds are arbitrary and controversial, a range of 0.5% to 1% sequence divergence is often used to delineate the species taxonomic rank (Clarridge, III, 2004). Sequencing the 16S gene is currently the most common approach used in microbial classification as a result of its phylogenetic properties and the large amount of 16S gene sequences available for comparison analyses.

16S gene sequence databases

Accurate identification of organisms by comparative analysis of 16S gene sequences is strongly dependent on the quality of the database used. The curated Ribosomal Database Project (RDP-II, http://rdp.cme.msu.edu/) provides 623,174 bacterial and archaeal small subunit rRNA gene sequences in an aligned and annotated format and has achieved major improvements in the detection of sequence anomalies (Cole et al, 2007). Notably, among all of the online tools provided by the RDP-II web site, the RDP classifier tool has demonstrated effective taxonomic classification of short sequences produced by the new pyrosequencing technology. The Greengenes project (http://greengenes.lbl.gov/) offers annotated, chimera-checked, full-length 16S gene sequences in standard alignment formats (DeSantis et al, 2006) and has a particularly useful tool for 16S microarray design. The Silva project (http://www.arb-silva.de) (Pruesse et al, 2007) provides SSU as well as large subunit (LSU) rRNA sequences from Bacteria, Archaea, and Eukarya in a format that is fully compatible with the ARB package (Ludwig et al, 2004). The ARB package (www.arb-home.de) has been used in major 16S surveys (Eckburg et al, 2005, Ley et al, 2005, Ley et al, 2006, McKenna et al, 2008, Turnbaugh et al, 2006) and notably allows phylogenetic tree constructions by insertion of partial or near-full sequences into a pre-established phylogenetic tree using a parsimony insertion tool.

However, the lack of quality control of sequence entries (ragged sequence ends and outdated or faulty entries) in these major public sequence databases has led to the development of high quality commercial databases, including MicroSeq 500 and the RIDOM Mycobacteria project

(http:/ /www.ridom-rdna.de) (Harmsen et al, 2002). The MicroSeq 500 database targets the first 527-bp fragment of the 16S gene and is able to identify most of the clinically important bacterial strains with ambiguous biochemical profiles (Woo et al, 2003). The ribosomal differentiation of the medical microorganism (RIDOM) database targets the 5'end of the 16S sequence and is dedicated to Mycobacteria family analyses. However, although these commercial databases are continually expanding, the current total number of 16S entries remains uncertain, and the representation of taxonomic divisions is limited.

MEASURES OF MICROBIAL DIVERSITY

The assessment of microbial diversity in a natural environment involves two aspects, species richness (number of species present in a sample) and species evenness (distribution of relative abundance of species) (Magurran, 2005). In order to estimate species richness, researchers widely rely on the assignment of 16S sequences into Operational Taxonomic Unit (OTU or phylotype) clusters, for instance, as performed by DOTUR (Schloss and Handelsman, 2005). The criterion used to define an OTU at the species level is the percentage of nucleotide sequence divergence; the cutoff values vary between 1%, 3%, or 5%, depending on the study. As a result of these inconsistencies, reliable statistical comparisons or descriptions of species richness across studies are restricted (Martin,

2002). The total community diversity of a single environment, or the α-diversity, is often represented by rarefaction curves. These curves plot the cumulative number of OTUs or phylotypes captured as a function of sampling effort and, therefore, indicate only the OTU richness observed in a given set of samples (Eckburg et al, 2005). In contrast, nonparametric methods, including Chao1 or ACE, are richness estimators of overall α-diversity (Roesch et al, 2007). In addition, quantitative methods such as the Shannon or Simpson indices measure the evenness of the α-diversity. However, although these estimators can describe the diversity of the microbiota associated with a healthy or diseased state, they are not informative of the (phylo)genetic diversity of an environmental sample (Martin, 2002).

Contrary to the α-diversity, the β-diversity measure offers a community structure comparison (taxon composition and relative abundance) between two or more environmental samples. For instance, β-diversity indices can compare similarities and differences in microbial communities in healthy and diseased states. A broad range of qualitative (presence/absence of taxa) and quantitative (taxon abundance) measures of community distance are available using several tools, including LIBHUFF, P-test, TreeClimber (Schloss and Handelsman, 2006b), SONS (Schloss and Handelsman, 2006a), DPCoA, or UniFrac (Lozupone et al, 2006, Lozupone and Knight, 2005); these methods have been thoroughly detailed in a previous review (Lozupone and Knight, 2008). For example, the robust unweighted UniFrac tool (Liu et al, 2007) measures the phylogenetic distance between two communities as the fraction of phylogenetic tree branch lengths leading to a descendant from either one community or the other. While UniFrac efficiently detects differences in the presence or absence of bacterial lineages, the recently developed weighted UniFrac is the qualitative version of original UniFrac and provides an efficient detection of differences in the relative abundance of bacterial lineages (Lozupone et al, 2007).

16S HIGH-THROUGHPUT METHODS

16S clonal sanger sequencing

Until the appearance in the two last decades of sequencing- by-synthesis methods (Ronaghi et al, 1996) such as those used in pyrosequencing, the

Sanger sequencing method (Sanger et al, 1977) was the cornerstone of DNA sequence production. The Sanger method is based on DNA synthesis on a single-stranded template and di-deoxy chain-termination (Hall, 2007, Pettersson et al, 2008). Improvements in cost-effectiveness and the development of high-throughput techniques (e.g, fluorescent-labeled terminators, capillary separation, template preparation) (Hunkapiller et al, 1991) have enabled direct sequencing of clones, without laborious prior screening by restriction analysis. As a result, Sanger sequencing has produced the earliest in-depth analyses of microbial communities (Eckburg et al, 2005, Ley et al, 2006, Turnbaugh et al, 2006, Zhou et al, 2004). All 16S genes in a sample are amplified using a universally conserved primer pair that targets most of the species that have been sequenced and deposited in the ribosomal databases. After the cloning of amplified PCR products into specific vectors, the inserts are sequenced. Sanger sequencing offers high phylogenetic resolution power, as the method yields the longest sequencing read available, up to 1000 base pairs (bp) (Shendure and Ji, 2008).

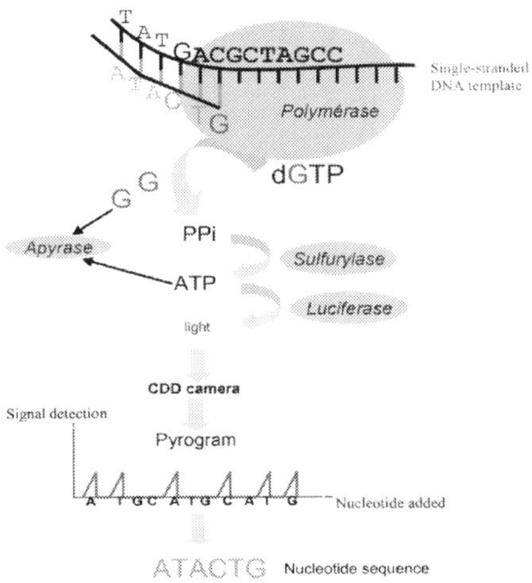

Figure 1: Principle of pyrosequencing technology. The primer for the sequencing step is hybridized to a single-stranded DNA template, and incubated with the enzymes, DNA polymerase, ATP sulfurylase, luciferase and apyrase, and the substrates. Deoxyribonucleotide triphosphate (dNTP) is added, one at a time, to the pyrosequencing reaction. The incorporation of a nucleotide is accompanied by release of pyrophosphate (PPi). The ATP sulfurylase quantitatively converts PPi to ATP. The signal light produced by the luciferase-catalyzed reaction in

presence of ATP is detected by a charge coupled device (CCD) camera and integrated as a peak in a Pyrogram. The nucleotide degrading Apyrase enzyme continuously degrades ATP excess and unincorporated dNTPs. The process continues with addition of the next dNTP and the nucleotide sequence of the complementary DNA strand is inferred from the signal peaks of the pyrogram.

16S microarray-based approach

The DNA microarray is a powerful technology that can simultaneously detect thousands of genes on a single glass slide or silicon surface (Gentry et al, 2006). Mainly used in gene expression profiling (DeRisi et al, 1997, Schena et al, 1996), DNA microarray technology has also been employed in bacterial identification and, more recently, has been adapted for exploring microbial community diversity in environmental niches. Microarrays used for microbial identification rely on the 16S gene and use short 20- to 70-mer oligonucleotide probes (Bae and Park, 2006) for multi-species detection. These are referred to as phylogenetic oligonucleotide microarrays (POAs).

Due to the phenomenal microbial diversity that might be present in an environmental sample and the lack of prior knowledge of its population composition, the oligonucleotide probe design strategy has been modified for use in community diversity analysis. Instead of using one unique probe that targets a specific region of one taxon, multiple probe sets are employed to target microorganisms at different taxonomic levels. An efficient probe design based on a hierarchical phylogenetic framework was established in 2003, using a database of curated and aligned 16S sequences known as ProkMSA (DeSantis et al, 2003). Desantis et al. defined 9,020 OTUs in the ProkMSA alignment using sequence identity clustering and demonstrated the capability of microarrays to correctly assign two species to their OTU using multiple probe sets (DeSantis et al, 2003).

High-throughput microorganism detection by microarray technology was first achieved with high-density photolithography microarrays using 31,179 20-mer oligonucleotide probes in a deep air investigation (Wilson et al, 2002). These high density microarrays present an alternative to clone library sequencing, since they can more deeply assess microbial diversity (DeSantis et al, 2007) without a cloning bias. With advancements in technology, the PhyloChip platform, an Affymetrix microarray product, was developed by the Lawrence Berkeley National Laboratory. The PhyloChip has rapidly identified up to 8,900 distinct environmental microorganisms at different taxonomy levels from soil (Yergeau et al, 2008) , air (Brodie et al, 2007), or uranium-contaminated site samples (Brodie et al, 2006) in a single experimental run. The PhyloChip is a glass slide with a small surface area, containing a high density microarray of hundreds of immobilized oligonucleotides (15- to 25-mer length). The method employs parallel hybridization reactions, using a flow of fluorescently labeled DNA targets. The microarray slide is analyzed with a fluorescence microscope equipped with a cooled CCD camera.

PYROSEQUENCING TECHNOLOGY

Sequencing chemistry

Pyrosequencing is a DNA sequencing method (Clarke, 2005, Ronaghi, 2001, Ronaghi and Elahi, 2002) based on the sequencing-by-synthesis principle, which was first described in 1985 (Melamede and Wallace, 1985). This method relies on efficient detection of the sequential incorporation of natural nucleotides during DNA synthesis (Ronaghi et al, 1996, Ronaghi et al, 1998) (Figure 1). The pyrosequencing technique includes four enzymes that are involved in a cascade reaction system (Figure 1).

During the reaction, the Klenow fragment of DNA polymerase I releases inorganic pyrophosphate molecules (PPi) upon the addition of one nucleotide to a primer hybridized to a single-stranded DNA template. The second reaction, catalyzed by ATP sulfurylase, produces ATP, using the released PPi as a substrate. The ATP molecules are then converted to a luminometric signal by the luciferase enzyme. Therefore, the signal light is detected only if a base pair is formed with the DNA template, and the signal strength is proportional to the number of nucleotides incorporated in a single nucleotide flow. Finally, the unincorporated nucleotides and excess ATP are degraded between base additions by a nucleotide-degrading enzyme such as apyrase (Ronaghi et al, 1998); at this point, another dNTP is added and a new cycle begins. The earliest attempts at pyrosequencing were performed using the PSQ96 system (Biotage AB, Upsala, Sweden) and targeted the short 16S variable regions V1 and

V3 (Jonasson et al, 2002, Tarnberg et al, 2002) or the human pathogen H. pylori (Hjalmarsson et al, 2004). This system produced reads of an average length of around 20- 40 bases.

The 454 life sciences pyrosequencing platform

In 2005, Margulies et al. first described a highly parallel sequencing platform (GS 20 454 Life Sciences) using a pyrosequencing protocol optimized for solid support. The authors demonstrated the ability of the system to assemble complete genomes (Mycoplasma genitalium and Streptococcus pneumoniae) from short sequencing reads (Margulies et al, 2005). In 454 pyrosequencing, the DNA template is fragmented, and the resulting fragments are individually immobilized onto a bead by limiting dilution. Emulsion PCR is performed for the DNA amplification step, in which each DNA fragment is independently confined into a droplet of oil and water containing PCR reagents. The clonally amplified fragments are distributed into a picotiter plate, which contains ~1.6 million picoliter wells, with a well diameter allowing one bead per well.

Using the pyrosequencing protocol previously described, the chemiluminescent signal obtained from an incorporated nucleotide is recorded by a charge-coupled device (CCD) camera, and data analysis, such as image processing or de novo sequence/genome assembly, is performed with provided bioinformatics tools. Other massively parallel platforms (Solexa and SOLID) using flow cell and other sequencing chemistries were reviewed previously (Holt and Jones, 2008).

While the first generation 454 Life Sciences apparatus (GS20) provided up to 25 megabases of data with an average read length of 100 base pairs (bp), the new GS FLX Titanium provides up to 400 megabases of data with an average read length of 400 bp. The 454 Life Sciences highthroughput sequencing platform outperforms the clonal Sanger sequencing method in terms of time and cost per sequenced nucleotide (Hugenholtz and Tyson, 2008), capacity of sequencing per run, as well as time savings in library preparation and data analysis. The contribution of 454 pyrosequencing to the literature is considerable, as 283 publications since 2005 have reported the use of this technology (Figure 2), and more than 20% of those publications belong to Science or Nature group journals. When applied to the field of microbial diversity, 454 pyrosequencing offers major insights, particularly in the investigation of human gut flora

(Dethlefsen et al, 2008, Turnbaugh et al, 2006, Turnbaugh et al, 2008), vaginal microbiota (Spear et al, 2008), hand surface bacteria (Fierer et al, 2008), soil diversity (Roesch et al, 2007), deep sea ecosystems (Huber et al, 2007, Sogin et al, 2006), and viral and phage populations (Desnues et al, 2008, Eriksson et al, 2008, La et al, 2008). Due to the short read length generated by the 454 platform and in order to increase sequencing capacity, new strategies for exploring microbial diversity by 16S pyrosequencing were developed.

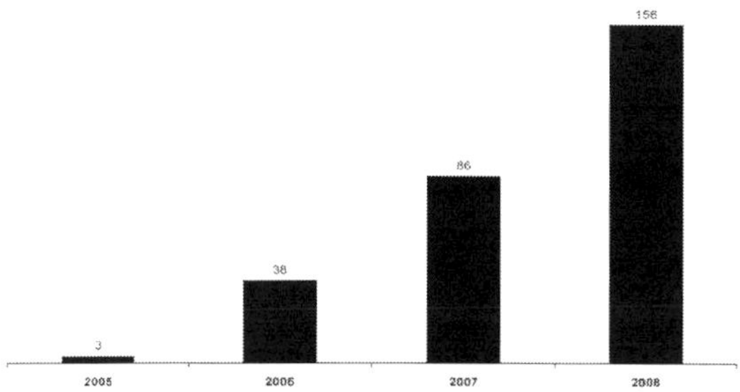

Figure 2: Number of publications enabled by the 454 pyrosequencing technology since 2005.

16S PYROSEQUENCING STRATEGIES

Targeting specific regions

Due to the short sequencing read length generated by pyrosequencing technology (e.g, 100 bp for GS 20 and 200 bp for GS FLX) and due to the small amount of nucleotide variability in the 16S gene throughout the bacterial domain, full 16S gene sequence assembly and the taxonomic assignment of species present in a mixed microbial sample remain a computational challenge (Armougom and Raoult, 2008). One strategy of addressing this problem consists of targeting a specific variable 16S region that exhibits a sufficient phylogenetic signal to be accurately assigned at the genus level or below. Surprisingly, short sequence fragments (including 100 bases) can provide substantial phylogenic resolution (Liu et al, 2007). By choosing appropriate 16S regions and simulating the read length obtained with GS FLX (250 bases), Liu et al. reproduced the same results obtained from full length 16S sequences using the UniFrac clustering tool (Liu et al, 2007). The authors suggested that the F8- R357 primer set, which amplifies a sequence spanning the V2 and V3 hypervariable regions and generates a 250-bp amplicon, could be the optimal primer set for exploring microbial diversity using 16S 454 pyrosequencing with GS FLX. These primers were also used in a study of the macaque gut microbiome (McKenna et al, 2008). A more recent study reported by Liu et al, focused on the capabilities of different taxonomic assignment methods (as a function of se- lected 16S regions), confirmed their previous conclusions recommending the F8-R357 primer set. The authors also suggested that regions surrounding the V6 hypervariable region were not appropriate for taxonomic assignment using a 16S 454 pyrosequencing strategy (Liu et al, 2008).

Whereas Liu et al. laud the use of the F8-R357 primer pair the study that introduced the concept of tag pyrosequencing was performed using the V6 hypervariable region amplified from deep water samples of the North Atlantic (Sogin et al, 2006). Based on the Shannon entropy measure, the 16S gene shows high variability in the V6 region. This region was selected in an analysis of the human gut microbiome combined with a barcoding strategy (Andersson et al, 2008). In addition, in their recent study of a gut microbial community, Huse et al. showed that the use of tags of the V3 and V6 regions in 454 pyrosequencing notably provides taxonomic assignments equivalent to those obtained by full length sequences generated using clonal Sanger sequencing (Huse et al, 2008).

The taxonomic assignment of pyrosequencing reads is sensitive to the classification method employed (Liu et al, 2008). Wang et al. showed a classification accuracy map of their RDP classifier tool along the 16S sequence position (Wang et al, 2007), which revealed that variable V2 and V4 regions of 16S provided the best taxonomic assignment results at the genus level. No true consensus seems to emerge among the 16S 454 pyrosequencing studies that target variable regions (Table1). However, this is not surprising since the power of phylogenetic resolution of variable 16S regions might differ depending on the taxa present in the microbial community studied; these results may also be due to the under-representation of 16S sequences of certain environments in the reference databases (Huse et al, 2008).

Table 1: Choice of primer set in microbial diversity investigations using 16S 454 pyrosequencing A Mixture of Forward (F) or Reverse (R) primers is indicated by "&". In italic, primer set targeting the Archaeal domain. Studies that used the barcoding strategy are indicated with an asterisk (*). The 16S rRNA variable region (Vx) targeted by the study is indicated in the last column.

Reference	Samples	Forward Primer	Reverse Primer	16S regions
(Roesch et al, 2007)	**Soil**	787F GNTACCTTGTTACGACTT	1492R ATTAGATACCCNGGTAG	**V5 to V9**
(Andersson et al, 2008)	Human gut	784F: GCCTTGCCAGCCCGCTCAG	1061R: AGGATTAGATACCCTGGTA	***V5-V6**
(Dowd et al, 2008a)	Cattle feces	530F: GTGCCAGCMGCNGCGG	1100R: GGGTTNCGNTCGTTG	***V4 to V6**
(Huse et al, 2008)	Human gut & Deap sea vent	338F: ACTCCTACGGGAGGCAGCAG 967F: CAACGCGAAGAACCTTACC & ATACGCGA|AG|GAACCTTACC	533R: TTACCGCGGCTGCTGGCAC 1046R: AGGTGNTGCATGGCTGTCG & 1046R1: AGGTGNTGCATGGTTGTCG	**V3** - **V6**
(McKenna et al, 2008)	Macaque gut	F8: AGAGTTTGATCCTGGCTCAG	R357: CTGCTGCCTYCCGTA	***V2-V3**
(Sogin et al, 2006)	Deep sea	967F: CAACGCGAAGAACCTTACC	1046R: CGACAGCCATGCANCACCT	**V6**
(Kim et al, 2008)	Tidal flat sediments	bact363F:CAATGGRSGVRASYCTGAHS *Arch339F: GGYGCASCAGGCGCGVAW*	Bact531R:CTNYGTMTTACCGCGGCTGC *Arch523R: TMCCGCGGCKGCTGVCASC*	**V3**
(Turnbaugh et al, 2008)	human Gut	*8F:* AGAGTTTGATCCTGGCTCAG F : See (Huber et al, 2007)	*338R:* RTGCTGCCTCCCGTAGGAGT R: See (Huber et al, 2007)	***V2** ***V6**
(Keijser et al, 2008)	Oral cavity	909F: AAACTYAAARRAATTGACGG & 917F: GAATTGACGGGGRCCCGCA	1061R: TCACGRCACGAGCTGACGAC	**V6**
(Fierer et al, 2008)	Hand surface	27F: AGAGTTTGATCCTGGCTCAG	338R: CATGCTGCCTCCCGTAGGAGT	***V2-V3**
(Dethlefsen et al, 2008)	Human Gut	F1: CAACGCGAAGAACCTTACC & F2: ATACGCGAGGAACCTTACC F: ACTCCTACGGGAGGCAGCAG	R: CGACARCCATGCASCACCT R: TTACCGCGGCTGCTGGCAC	***V6** ***V3**
(Spear et al, 2008)	vagina	27F: AGAGTTTGATCCTGGCTCAG	355R: GCTGCCTCCCGTAGGAGT	***V2-V3**
(Huber et al, 2007)	Deep sea	967F-PP: CNACGCGAAGAACCTTANC & 967F-UC1: CAACGCGAAAAACCTTACC & 967F-UC2: CAACGCGCAGAACCTTACC & 967F-UC3: ATACGCGARGAACCTTACC & 967F-AQ: CTAACCGANGAACCTYACC *958arcF: AATTGGANTCAACGCCGG*	1046R CGACAGCCATGCANCACCT& 1046RPP CGACAACCATGCANCACCT& 1046RAQ1 CGACGGCCATGCANCACCT& 1046RAQ2 CGACGACCATGCANCACCT *1048arcR: CGRCGGCCATGCACCWC & 1048arcR: CGRCRGCCATGYACCWC*	***V6**

16S barcoding strategy

The currently available 454 GS FLX pyrosequencer can accommodate a maximum of sixteen independent samples, since a picotiter plate contains sixteen physically separated regions.

To overcome this limitation and expand the capacity by pooling DNA from independent samples in a single sequencing run, a barcoding approach was developed, which associates a short unique DNA sequence tag (barcode) with each DNA template origin. In contrast with Sanger sequencing, pyrosequencing technology such as GS FLX generates sequencing reads from the first position of the DNA template fragment. Therefore, the sequencing reaction driven by an oligonucleotide that is complementary to adaptor A and B first reads the barcode sequence, allowing the identification of the original DNA template source (Andersson et al, 2008, McKenna et al, 2008). Using the DNA barcode strategy, each primer is composed of the 5'-adaptor A or B (required for the PCR emulsion), followed by the DNA barcode and the primer targeting the DNA region (McKenna et al, 2008) (Figure 3). By selecting a DNA target with a nucleotide length inferior to the average read length of pyrosequencing, an unidirectional barcoding strategy can be achieved, adding the barcode sequence only to the forward or reverse primer (Andersson et al, 2008). The length of the barcode sequence varies from 2-4 nucleotides (Andersson et al, 2008, Binladen et al, 2007, McKenna et al, 2008) to 10 nucleotides (Dowd et al, 2008b, Dowd et al, 2008a) or 12 nucleotides (Fierer et al, 2008, Turnbaugh et al, 2008); the number of DNA samples that can be combined in a single sequencing run increases with barcode length.

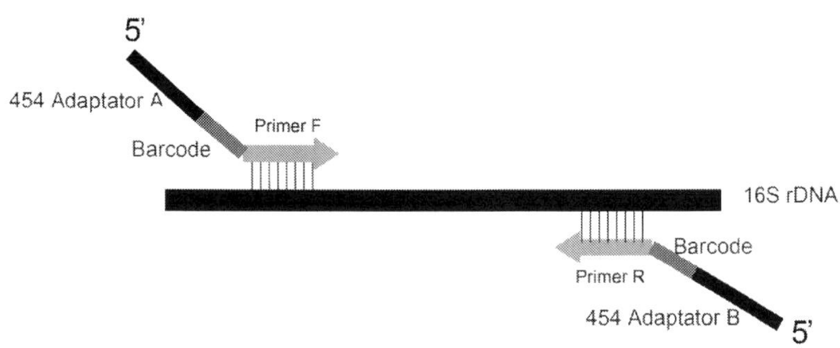

Figure 3: The barcoding strategy used in 454 pyrosequencing Each primer contains a region complementary to the 454 A or B adaptator sequencing primers, a variable barcode sequence length (in red) and a region complementary to ends of the 16S rRNA region targeted (Primer R or F)

Finally, the efficiency of barcoding strategy was investigated using a set of eight nucleotide barcodes based on error- correcting codes (called Hamming codes). The ability to detect sequencing errors that change sample assignments and to correct errors in the barcodes was evaluated to 92% (Hammady et al, 2008).

Taxonomy assignment of short pyrosequencing reads

Taxonomy assignment using standard phylogenic methods such as likelihood- or parsimony- based tree construction is inconceivable, given the large amount of sequence data (400,000 reads for GS FLX Titanium) generated by high-throughput pyrosequencing (Liu et al, 2008). In addition, due to the short read length resulting from pyrosequencing (~ 100 bp for GS 20 and 250 bp for GS FLX), full 16S sequence assembly is a computational challenge, especially for closely related species in a mixed bacterial sample. Therefore, current taxonomy classification tools such as the naïve Bayesian RDP-II classifier (Wang et al, 2007) or the Greengenes classifier (DeSantis et al, 2006) employ rapid but approximate methods. In contrast to the RDP-II classifier, the Greengenes classifier requires a pre-computed alignment for its taxonomic classification. While, tree-based methods are subject to large variations in assignment accuracy according to the DNA region examined (Liu et al, 2008), the tree-independent methods used in the RDP-II and Greengene classification tools yield stable and accurate taxonomy assignment results. However, although these methods provide satisfactory assignment results at the genus level, they have a limited resolution power at the species level. This limitation could be reduced, but not yet solved, by the 400- to 500-base read length generated by the new GS Titanium pyrosequencing platform.

Pyrosequencing read classification based on a sequence similarity search using the BLAST algorithm can yield reliable results; however, the closest match is not inevitably the nearest phylogenetic neighbor (Koski and Golding, 2001). Sundquist et al. (Sundquist et al, 2007) recently proposed a method based on a similarity search with BLAT, a BLAST like alignment tool (Kent, 2002), and inferred specific phylogeny placement using a preliminary set of best BLAT match scores. Finally, a tag-mapping methodology has been recently introduced with GAST (Global Alignment for Sequence Taxonomy), which is a taxonomic assignment tool combining BLAST, multiple sequence alignment, and global distance measures. GAST has been used to identify taxa present in deep sea vent and human gut samples (Huse et al, 2008, Sogin et al, 2006).

The accuracy rate of classification methods of short 16S rRNA sequences is measured as the percentage of sequences correctly classified from a representative set of bacteria sequences of known classification. This accuracy rate can be associated with the misclassified sequence rate. Overall, the simulation tests showed efficient classification down to the genus level. Simulating 200-base segments (such as generated by the GS FLX), the RPD-II classifier tool indicated an overall taxonomic assignment accuracy of 83.2 % at the genus level (Wang et al, 2007). Likewise, using the methodology of Sundquist and co-workers (Sundquist et al. 2007), the simulation of read resolution for 200 base segments in diverse and representative samples yield an accuracy rate around 80% as obtained for the RDP-II classifier. However, while the benchmark of the Sundquist method identified the V1 and the V2 regions of the 16S as the best targets for the pyrosequencing of 100 bases (such as generated by the GS 20), the benchmark of the RDP-II classifier method identified the V2 and the V4 regions. Finally, the comparison of the classification methods is rather difficult since the representative bacteria sequences of known classification used for the benchmarking is different for each method. A collection of reference sequences of known classification (defined as gold standard) is required for classification method comparisons. For instance, collections of reference sequence alignments are generally used for the benchmarking of multiple sequence alignment methods (Armougom et al. 2006).

Metagenomic approach

Metagenomic is a culture-independent genomic analysis of entire microbial communities inhabiting a particular niche, such as the human gut (Riesenfeld et al, 2004, Schmeisser et al, 2007). Metagenomic investigations aimed at finding "who's there and what are they doing" (Board On Life Sciences, 2007) are providing new insight into the genetic variability and metabolic capabilities of unknown or uncultured microorganisms (Turnbaugh et al, 2006). The inclusion of the metagenomic approach in this review was required, since metagenomic studies include the analysis of the 16S sequences contained in metagenome data, in order to identify community composition and determine bacterial relative abundance (Biddle et al, 2008, Edwards et al, 2006 Krause et al, 2008, Wegley et al, 2007). However, we distinguished the metagenomic approach from 16S highthroughput methods, as the principal purpose of metagenomic is to explore the entire gene content of metagenomes for metabolic pathways and to understand microbial community interactions (Tringe et al, 2005), rather than targeting a single gene such as 16S. In contrast with some scientific literature examples and because it can only answer "who is there", 454 pyrosequencing investigations based on the 16S gene should not be assimilated to a metagenomic case.

Before the commercialization of the 454 pyrosequencing platform, microbial community sequencing in early metagenomic studies, such as in the Sargasso sea (Venter et al, 2004) or in acid mine drainage (Edwards et al, 2000), involved preliminary clone library construction and capillary sequencing. This approach limited the expansion of microbial diversity knowledge due to cloning bias and the cost of capillary sequencing. In contrast, the inexpensive next-generation 454 pyrosequencing technology can perform direct sequencing without requiring preliminary PCR amplification or library construction. By excluding cloning and PCR bias, 454 sequencing revolutionized the metagenomic field by capturing up to 100% (depending on the quality of DNA extraction and environmental sampling) of the microbial diversity present in a sample.

CASE STUDIES

The Human microbiome project

Launched by the NIH Roadmap for Medical Research, the Human Microbiome Project (HMP) seeks to comprehensively characterize the human microbiota in order to better understand its role in human health and disease states (http://nihroadmap.nih.gov/) The HMP project mainly focuses on the gastrointestinal, oral, vaginal, and skin microbiota.

The gastrointestinal microbiota

Until recently, characterization of the gut microbiota diversity was restricted to culture-based methods (Simon and Gorbach, 1986). While the cultivable fraction is currently estimated to be 442 bacterial, 3 archaeal, and 17 eukaryotic species (Zoetendal et al, 2008), the species richness of the gut microbiota is estimated to be 15,000 or 36,000 bacterial species, depending on the similarity cut-off applied in OTU clustering (Frank et al, 2007). With the development of culture- independent methods, 16S gene surveys have deeply enhanced the microbial diversity map of the human gut microbiota (Eckburg et al, 2005, Gill et al, 2006, Ley et al, 2006, Palmer et al, 2007). A major large-scale 16S investigation by Eckburg et al. indicated that the gut of healthy individuals was mainly composed of the Bacteroidetes and Firmicutes bacterial phyla (83% of sequences). The sequences also included the archaeal Metanobrevibacter species, as well as a majority of uncultivated species and novel microorganisms (Eckburg et al, 2005). Surprisingly, at the phylum level, the bacterial diversity of the gut microbiota is low; only 8 of 70 known bacterial phyla are represented. Despite the predominance of Firmicutes and Bacteroidetes phylotypes, the gut microbiota displays a great inter-individual specificity in its composition (Ley et al, 2006), especially in newborn babies (Palmer et al, 2007). Within elderly populations, the Bacteroidetes proportion can decline (Woodmansey, 2007). Recently, a barcoded pyrosequencing study of the gut microbiota of six elderly individuals showed that Actinobacteria was the second most abundant phylum, not Bacteroidetes as expected(Andersson et al, 2008). Age, caloric intake, antibiotic treatment (Dethlefsen et al, 2008) and diet are a few environmental factors that can influence the gut microbiota composition and thus affect human health.

Through the comparison of lean and obese individuals, a possible relationship between obesity and the composition of (and changes in) gut microbiota has been investigated (Ley et al, 2006, Turnbaugh et al, 2008) and reviewed (Dibaise et al, 2008). Ley et al. showed that obese subjects have a higher Firmicutes/Bacteroidetes ratio than lean controls (Ley et al, 2006). By testing dietary factors, the authors demonstrated a shift in Bacteroidetes and Firmicutes relative abundance that correlated with weight loss. Reduced bacterial diversity and familial similarity of gut microbiota composition within obese individuals have also been recently reported (Turnbaugh et al, 2008).

The oral cavity microbiota

The understanding of healthy oral cavity microflora is essential for the prevention of oral diseases and requires unambiguous identification of microorganism(s) associated with pathology. For instance, it is accepted that Streptococcus mutans is the etiological agent in dental caries (Jenkinson and Lamont, 2005). Limited by the cloning and sequencing approach, the characterization of the diversity in human oral microflora was radically enhanced by the first 16S 454 pyrosequencing of saliva and supragingival plaque (Keijser et al, 2008). Keijer et al. revealed 3,621 and 6,888 phylotypes in saliva and plaque samples, respectively, and estimated the total microbial species richness to be 19,000 (3% similarity cut-off). Within the 22 phyla identified, the main taxa are Firmicutes (genus Streptococcus and Veillonella) and Bacteroidetes (genus Prevotella) in saliva, while Firmicutes and Actinobacteria (genus Corynebacterium and Actinomyces) are the most common in plaque.

The vaginal microbiota

By its exposure to the external environment, the female genital tract can be easily affected in its reproductive functions. Previous surveys of the human vaginal microbiota proposed that the normal vaginal microbiota can act as a defense mechanism, playing an essential role in preventing infections such as bacterial vaginosis or sexually transmitted diseases in women. Zhou et al. first characterized the vaginal microbiota by 16S clone library sequencing (Zhou et al, 2004). As found in culture-dependent studies, the authors showed that Lactobacilli and Atopobium were generally the predominant organisms; they also reported the first identification of a Megasphaera species in the vagina. By employing a full 16S pyrosequencing strategy and developing a classification method of short 16S pyrosequencing reads, Sundquist et al. studied the human vagina during pregnancy and corroborated previous results. The authors identified Lactobacillus as the dominant genus and detected a significant presence of other genera including Psychrobacter, Magnetobacterium, Prevotella, Bifidobacterium, and Veillonella (Sundquist et al, 2007). However, Lactobacillus can be missing in healthy vaginal microbiota and replaced by other predominant genera such as Gardnerella, Pseudomonas, or Streptococcus (Hyman et al, 2005). Although inter-individual variability of the vaginal microbiota was demonstrated, the function of the communities was conserved and was shown to be involved in the production of lactic acid (Zhou et al, 2004). The maintenance of the vaginal acidity preserves an unfriendly environment for the growth of many pathogenic organisms (Zhou et al, 2004). Important shifts in relative abundances and types of bacteria, especially the decrease in lactic acid bacteria, in the healthy vagina are associated with bacterial vaginosis infections (Spiegel, 1991). Compared to the healthy vaginal microbiota, bacterial vaginosis-associated microbiota showed greater specie richness, different bacterial community structures, and a strong association with members of the Bacteroidetes and Actinobacteria phyla (Oakley et al, 2008). To enhance the understanding of the important variations in the incidence of bacterial vaginosis among racial or ethnic groups, the vaginal microbiota of Caucasian and black women were explored. Striking differences were demonstrated in community abundance and also in composition; for instance, the predominance of Lactobacillus in black women is lower than in Caucasian women (Zhou et al, 2007).

The skin microbiota

The skin probably offers one of the largest human-associated habitats and has a bacterial density of around 107 cells per square centimeter (Fredricks, 2001). The commensal bacterial communities and pathogenic microorganisms harbored by the skin's surface suggest an association with healthy, infectious or noninfectious (psoriasis, eczema) (Grice et al, 2008) skin states. Until recently, because skin infections generally involve rare pathogenic isolates, the resident skin bacteria remained poorly described and limited to culture-dependent studies that under-represented the extent of bacterial diversity.

A recent 16S survey of the resident skin microbiota of the inner elbow region, an area subjected to atopic dermatitis, from five healthy humans generated 5,373 nearly complete 16S rRNA sequences. These sequences were assigned into 113 phylotypes belonging to the Proteobacteria (49%), Actinobacteria (28%), Firmicutes (12%), Bacteroidetes (9.7%), Cyanobacteria (<1%), and Acidobacteria(<1%) phyla (Grice et al, 2008). Most of the 16S sequences (90%) belong to the Proteobacteria phylum and, more accurately, to the Pseudomonas and Janthinobacterium genera. Finally, the authors' results indicated a low level of deep evolutionary lineage diversity and a similar diversity profile for all of the subjects, suggesting a common core skin microbiota among healthy subjects (Grice et al, 2008). Interestingly, the few cultivated commensal skin bacteria, including S. epidermidis and P. acnes, accounted for only 5% of the microbiota captured in the Grice et al. study (Grice et al, 2008).

In contrast, a study examining the diversity of skin microbiota from the superficial forearms in six healthy subjects indicated a small core set of phylotypes (2.2%) and a high degree of inter-individual variability in the microbiota (Gao et al, 2007). The distribution of the 182 identified phylotypes at the phylum level was 29% Proteobacteria, 35% Actinobacteria, 24% Firmicutes, 8% Bacteroidetes, 1.6% Deinococcus-Thermus, 0.5% Termomicrobia, and 0.5% Cyanobacteria (Gao et al, 2007). However, only the three most abundant phyla were observed in all subjects. Although many phylotypes overlap between the inner elbow and forearm skin microbiota, the predominant phylotype belonging to the Proteobacteria phyla of the inner elbow microbiota is missing in the forearm microbiota. In addition, the forearm skin microbiota possesses more members of the Actinobacteria and Firmicutes phyla (Grice et al, 2008). It is difficult, however, to compare studies that employed different methods and skin sample locations.

The characterization of skin microbial communities and their interactions is still in its infancy, since the impacts of sex, age, clothing, and others factors have not been clearly determined. However, a recent study on the reduction of disease transmission by hand washing used a barcoded pyrosequencing approach to characterize the hand surface microbiota of 51 healthy men and women to determine how specific factors could affect the community structure (Fierer et al, 2008). Although the authors detected a core set of bacterial taxa on the hand surface, the results mainly revealed a high intra- and inter-individual variation in community structure when sex, hand washing, or handedness factors were considered. In addition, though the diversity observed in hand surfaces is high (sequences from>25 phyla were identified), 94% of the sequences belong only to three of these phyla (Actinobacteria, Firmicutes and Proteobacteria) (Fierer et al, 2008). Finally, independently of the skin site sampled (Fierer et al, 2008, Gao et al, 2007, Grice et al, 2008), all of the studies shared the same predominant phyla: Proteobacteria, Actinobacteria, and Firmicutes.

LIMITS OF 16S ANALYSES

The 16S is an efficient phylogenic marker for bacteria identification and microbial community analyses. However, the multiple pitfalls of PCR-based analyses, including sample collection, cell lysis, PCR amplification, and cloning, can affect the estimation of the community composition in mixed microbial samples (Farrelly et al, 1995, von et al, 1997).

Although one or two 16S gene copies are commonly exhibited by a single genome, multiple and heterogeneous 16S genes in a single microbial genome are not rare and can lead to an overestimation of the abundance and bacterial diversity using culture-independent approaches (Acinas et al, 2004). Multiple copies of rRNA operons (rrn) per genome are generally found in rapidly growing microorganisms, especially in soil bacteria (Klappenbach et al, 2000). As a result, the number of copies of the 16S gene in a microbial genome can reach 10 or 12 copies in Bacillus subtilis (Stewart et al, 1982) or Bacillus cereus (Johansen et al, 1996), respectively, and up to 15 copies in Clostridium paradoxum (Rainey et al, 1996). In addition to the heterogeneity of the 16S gene copy number per genome, a bacterial species can display important nucleotide sequence variability among its 16S genes (Acinas et al, 2004, Rainey et al, 1996, Turova et al, 2001). Furthermore, it is well known that 16S gene sequencing lacks taxonomic resolution at the species level for some closely related species (Janda and Abbott, 2007), subspecies, or recently diverged species (Fox et al, 1992). In this way, Escherichia coli and E. fergusonii species, as well as the subspecies Bartonella vinsonii subsp. arupensis and B. vinsonii subsp. vinsonii, are indistinguishable, when using 16S sequence similarity comparisons (Adekambi et al, 2003).

Likewise, PCR amplification bias in a mixed microbial sample causes the taxon amplicon abundance to differ from the real proportions present in the community. Notably, PCR amplification bias can be induced by the choice of primer set, the number of replication cycles, or the enzyme system used (Qiu et al, 2001, Suzuki and Giovannoni, 1996). An obvious example is that the sensitivity of universal primers is limited to the currently known 16S sequences and does not reach 100% coverage. What is the primer sensitivity for unknown microbial sequences, and how can the same hybridization efficiency be guaranteed for all targets in the sample? However, metagenomic studies, which can theoretically access up to 100% of the microbial diversity in a sample, have yielded powerful alternatives to bypass primer and cloning bias using 454 pyrosequencing.

Another aspect limiting the capture of the true microbial diversity in a mixed sample by 16S surveys is inherent to the cloning step. The efficiency of ligation to the plasmid, transformation, and amplification in the host can all have an effect. For instance, it has been suggested that many unclonable genes in the E. coli host are present in a single copy per genome and hence are under-represented in clone libraries due to inactive promoters or toxic effects induced by gene transcription/translation into the host (Kunin et al,2008, Sorek et al, 2007). Finally, horizontal gene transfer and recombination events have also been reported in ribosomal genes (Miller et al, 2005), which distort phylogenetic signals and thereby affect the phylogenetic classification.

CONCLUSIONS AND PERSPECTIVES

Culture-independent methods based on the 16S rRNA gene yield a useful framework for exploring microbial diversity, by establishing the taxonomic composition and/or structure present in environmental samples using both a and b-diversity measures, phylogenetic tree construction, and sequence similarity comparison.

Unlikely to culture methods, these recent high-throughput methods allow accessing to the true microbial diversity. In a point of view of clinical research, new or uncultured etiologic agents from poly-microbial samples (pulmonary infections, brain abscess) of disease states can be identified and will lead to elaborate more appropriate antibiotherapies rather than the use of broad range antibiotics. In addition, compared to lean controls, the reduction of the Bacteroidetes members and the increase of the methanogen Methanobrevibacter smithii in obese patients were revealed by high-throughput sequencing methods. These results suggested that modulate the relative abundance of some microbial groups of the gut microbiota could be beneficial for obese treatment. The huge amount of sequences provided by these new sequencing methods hugely increase the number of 16S sequences in databases, and thus improve the ability of 16S sequence identification using sequence similarity search tools. In a near future, the accuracy of classification methods of short 16S sequences will be improved by the increase of read length (450pb) produced by the new 454 FLX Titanium apparatus. In addition, the increase in sequence production capabilities of the 454 FLX Titanium associated with the barcoding strategy will allow examining much more different samples in a single pyrosequencing run.

However, 16S high-throughput methods can not characterize the functional component (defined as the microbiome) of an environmental sample. Such limitations arise by targeting a sole gene marker. This limitation can be overcome by a metagenomic approach, which focuses on the full gene content (gene-centric analysis) of a sample. Therefore, in addition to providing species richness and evenness information, the relatively unbiased metagenomic approach can also identify the metabolic capabilities of a microbial community and disclose specific adaptive gene sets that are potentially beneficial for survival in a given habitat.

REFERENCES

1. Acinas SG, Marcelino LA, Klepac CV, Polz MF (2004) Divergence and redundancy of 16S rRNA sequences in genomes with multiple rrn operons. J Bacteriol 186: 2629- 2635.
2. Adekambi T, Colson P, Drancourt M (2003) RpoB-based identification of nonpigmented and late-pigmenting rapidly growing mycobacteria. J Clin Microbiol 41: 5699- 5708.
3. Andersson AF, Lindberg M, Jakobsson H, Backhed F, Nyren P, et al. (2008) Comparative analysis of human gut microbiota by barcoded pyrosequencing. PLoS ONE 3: e2836.
4. Armougom F, Raoult D (2008) Use of pyrosequencing and DNA barcodes to monitor variations in Firmicutes and Bacteroidetes communities in the gut microbiota of obese humans. BMC Genomics 9: 576.
5. Armougom F, Moretti S, Keduas V, Notredame C (2006) The iRMSD: a local measure of sequence alignment accuracy using structural information. Bioinformatics 22: e35-e39.
6. Bae JW, Park YH (2006) Homogeneous versus heterogeneous probes for microbial ecological microarrays. Trends Biotechnol 24: 318-323.
7. Biddle JF, Fitz GS, Schuster SC, Brenchley JE, House CH (2008)

Metagenomic signatures of the Peru Margin subseafloor biosphere show a genetically distinct environment. Proc Natl Acad Sci USA 105: 10583-10588.

8. Binladen J, Gilbert MT, Bollback JP, Panitz F, Bendixen C, et al. (2007) The use of coded PCR primers enables high-throughput sequencing of multiple homolog amplification products by 454 parallel sequencing. PLoS ONE 2: e197.
9. Board On Life Sciences (2007) The New Science of Metagenomics: Revealing the Secrets of Our Microbial Planet. NATIONAL RESEARCH COUNCIL OF THE NATIONAL ACADEMIES ed.
10. Brodie EL, DeSantis TZ, Joyner DC, Baek SM, Larsen JT, et al. (2006) Application of a high-density oligonucleotide microarray approach to study bacterial population dynamics during uranium reduction and reoxidation. Appl. Environ. Microbiol 72: 6288-6298.
11. Brodie EL, DeSantis TZ, Parker JP, Zubietta IX, Piceno YM, et al. (2007) Urban aerosols harbor diverse and dynamic bacterial populations. Proc Natl Acad Sci USA 104: 299-304.
12. Clarke SC (2005) Pyrosequencing: nucleotide sequencing technology with bacterial genotyping applications. Expert Rev Mol Diagn 5: 947-953.
13. Clarridge JE III (2004) Impact of 16S rRNA gene sequence analysis for identification of bacteria on clinical microbiology and infectious diseases. Clin Microbiol Rev17: 840-62.
14. Cole JR, Chai B, Farris RJ, Wang Q, Kulam SMAS, et al. (2007) The ribosomal database project (RDP-II): introducing myRDP space and quality controlled public data. Nucleic Acids Res 35: D169-D172.
15. DeRisi JL, Iyer VR, Brown PO (1997) Exploring the metabolic and genetic control of gene expression on a genomic scale. Science 278: 680-686.
16. DeSantis TZ, Brodie EL, Moberg JP, Zubieta IX, Piceno YM, et al. (2007) High-density universal 16S rRNA microarray analysis reveals broader diversity than typical clone library when sampling the environment. Microb Ecol 53: 371-383.
17. DeSantis TZ, Dubosarskiy I, Murray SR, Andersen GL (2003) Comprehensive aligned sequence construction for automated design of effective probes (CASCADE-P) using 16S rDNA. Bioinformatics

19: 1461-1468.

18. DeSantis TZ, Hugenholtz P, Larsen N, Rojas M, Brodie EL, et al. (2006) Greengenes, a chimera-checked 16S rRNA gene database and workbench compatible with ARB. Appl Environ Microbiol 72: 5069-5072.
19. Desnues C, Rodriguez BB, Rayhawk S, Kelley S, Tran T, et al. (2008) Biodiversity and biogeography of phages in modern stromatolites and thrombolites. Nature 452: 340-343.
20. Dethlefsen L, Huse S, Sogin ML, Relman DA (2008) The Pervasive Effects of an Antibiotic on the Human Gut Microbiota, as Revealed by Deep 16S rRNA Sequencing. PLoS Biol 6: e280.
21. Dibaise JK, Zhang H, Crowell MD, Krajmalnik BR, Decker GA, et al. (2008) Gut microbiota and its possible relationship with obesity. Mayo Clin Proc 83: 460-469.
22. Dowd SE, Callaway TR, Wolcott RD, Sun Y, McKeehan T, et al. (2008a) Evaluation of the bacterial diversity in the feces of cattle using 16S rDNA bacterial tag-encoded FLX amplicon pyrosequencing (bTEFAP). BMC Microbiol 8: 125.
23. Dowd SE, Sun Y, Wolcott RD, Domingo A, Carroll JA (2008b) Bacterial tag-encoded FLX amplicon pyrosequencing (bTEFAP) for microbiome studies: bacterial diversity in the ileum of newly weaned Salmonellainfected pigs. Foodborne Pathog Dis 5: 459-472.
24. Eckburg PB, Bik EM, Bernstein CN, Purdom E, Dethlefsen L, et al. (2005) Diversity of the human intestinal microbial flora. Science 308: 1635-1638.
25. Edwards KJ, Bond PL, Gihring TM, Banfield JF (2000) An archaeal iron-oxidizing extreme acidophile important in acid mine drainage. Science 287: 1796-1799.
26. Edwards RA, Rodriguez BB, Wegley L, Haynes M, Breitbart M, et al. (2006) Using pyrosequencing to shed light on deep mine microbial ecology. BMC Genomics 7: 57.
27. Eriksson N, Pachter L, Mitsuya Y, Rhee SY, Wang C, et al. (2008) Viral population estimation using pyrosequencing. PLoS Comput Biol 4: e1000074.
28. Farrelly V, Rainey FA, Stackebrandt E (1995) Effect of genome

size and rrn gene copy number on PCR amplification of 16S rRNA genes from a mixture of bacterial species. Appl Environ Microbiol 61: 2798-2801.

29. Fierer N, Hamady M, Lauber CL, Knight R (2008) The influence of sex, handedness, and washing on the diversity of hand surface bacteria. Proc Natl Acad Sci USA 105: 17994-17999.
30. Fox GE, Wisotzkey JD, Jurtshuk P Jr (1992) How close is close: 16S rRNA sequence identity may not be sufficient to guarantee species identity. Int J Syst Bacteriol 42: 166-170.
31. Frank DN, St Amand AL, Feldman RA, Boedeker EC, Harpaz N, et al. (2007) Molecular-phylogenetic characterization of microbial community imbalances in human inflammatory bowel diseases. Proc Natl Acad Sci USA 104: 13780-13785.
32. Fredricks DN (2001) Microbial ecology of human skin in health and disease. J Investig Dermatol Symp Proc 6:167-169.
33. Gao Z, Tseng CH, Pei Z, Blaser MJ (2007) Molecular analysis of human forearm superficial skin bacterial biota. Proc Natl Acad Sci USA 104: 2927-2932.
34. Gentry TJ, Wickham GS, Schadt CW, He Z, Zhou J (2006) Microarray applications in microbial ecology research. Microb Ecol 52: 159-175.
35. Gill SR, Pop M, Deboy RT, Eckburg PB, Turnbaugh PJ, et al. (2006) Metagenomic analysis of the human distal gut microbiome. Science 312: 1355-1359.
36. Giovannoni SJ, Stingl U (2005) Molecular diversity and ecology of microbial plankton. Nature 437: 343-348.
37. Grice EA, Kong HH, Renaud G, Young AC, Bouffard GG, et al. (2008) A diversity profile of the human skin microbiota. Genome Res 18: 1043-1050.
38. Hall N (2007) Advanced sequencing technologies and their wider impact in microbiology. J Exp Biol 210: 1518- 1525.
39. Hamady M, Walker JJ, Harris JK, Gold NJ, Knight R (2008) Error-correcting barcoded primers for pyrosequencing hundreds of samples in multiplex. Nat Methods 5: 235-237.
40. Harmsen D, Rothganger J, Frosch M, Albert J (2002) RIDOM:

Ribosomal Differentiation of Medical Microorganisms Database. Nucleic Acids Res 30: 416-417.

41. Hjalmarsson S, Alderborn A, Fock C, Muldin I, Kling H, et al. (2004) Rapid combined characterization of microorganism and host genotypes using a single technology. Helicobacter 9: 138-145.
42. Holt RA, Jones SJ (2008) The new paradigm of flow cell sequencing. Genome Res 18: 839-846.
43. Huber JA, Mark WDB, Morrison HG, Huse SM, Neal PR, et al. (2007) Microbial population structures in the deep marine biosphere. Science 318: 97-100.
44. Hugenholtz P, Tyson GW (2008) Microbiology: metagenomics. Nature 455: 481-483.
45. Hunkapiller T, Kaiser RJ, Koop BF, Hood L (1991)Large-scale and automated DNA sequence determination. Science 254: 59-67.
46. Huse SM, Dethlefsen L, Huber JA, Welch DM, Relman DA, et al. (2008) Exploring microbial diversity and taxonomy using SSU rRNA hypervariable tag sequencing. PLoS Genet 4: e1000255.
47. Hyman RW, Fukushima M, Diamond L, Kumm J, Giudice LC, et al. (2005) Microbes on the human vaginal epithelium. Proc Natl Acad Sci USA 102: 7952-7957.
48. Janda JM, Abbott SL (2007) 16S rRNA gene sequencing for bacterial identification in the diagnostic laboratory: pluses, perils, and pitfalls. J Clin Microbiol 45: 2761- 2764.
49. Jenkinson HF, Lamont RJ (2005) Oral microbial communities in sickness and in health. Trends Microbiol13: 589-595.
50. Johansen T, Carlson CR, Kolsto AB (1996) Variable numbers of rRNA gene operons in Bacillus cereus strains. FEMS Microbiol Lett 136: 325-328.
51. Jonasson J, Olofsson M, Monstein HJ (2002) Classification, identification and subtyping of bacteria based on pyrosequencing and signature matching of 16S rDNA fragments. APMIS 110: 263-272.
52. Keijser BJ, Zaura E, Huse SM, van d V, Schuren FH, et al. (2008) Pyrosequencing analysis of the oral microflora of healthy adults. J Dent Res 87: 1016-1020.

53. Kent WJ (2002) BLAT-the BLAST-like alignment tool. Genome Res 12: 656-664.
54. Kim BS, Kim BK, Lee JH, Kim M, Lim YW, et al. (2008) Rapid phylogenetic dissection of prokaryotic community structure in tidal flat using pyrosequencing. J Microbiol 46: 357-363.
55. Klappenbach JA, Dunbar JM, Schmidt TM (2000) rRNA operon copy number reflects ecological strategies of bacteria. Appl Environ Microbiol 66: 1328-1333.
56. Koski LB, Golding GB (2001) The closest BLAST hit is often not the nearest neighbor. J Mol Evol 52: 540-542.
57. Krause L, Diaz NN, Edwards RA, Gartemann KH, Kromeke H, et al. (2008) Taxonomic composition and gene content of a methane-producing microbial community isolated from a biogas reactor. J Biotechnol 136: 91-101.
58. Kunin V, Copeland A, Lapidus A, Mavromatis K, Hugenholtz P (2008) A bioinformatician's guide to metagenomics. Microbiol. Mol Biol Rev 72: 557-78.
59. La SB, Desnues C, Pagnier I, Robert C, Barrassi L, et al. (2008) The virophage as a unique parasite of the giant mimivirus. Nature 455: 100-104.
60. Lane DJ, Stahl DA, Olsen GJ, Heller DJ, Pace NR (1985) Phylogenetic analysis of the genera Thiobacillus and Thiomicrospira by 5S rRNA sequences. J Bacteriol 163: 75-81.
61. Ley RE, Backhed F, Turnbaugh P, Lozupone CA, Knight RD, et al. (2005) Obesity alters gut microbial ecology. Proc Natl Acad Sci USA 102: 11070-11075.
62. Ley RE, Turnbaugh PJ, Klein S, Gordon JI (2006) Microbial ecology: human gut microbes associated with obesity. Nature 444: 1022-1023.
63. Liu Z, DeSantis TZ, Andersen GL, Knight R (2008) Accurate taxonomy assignments from 16S rRNA sequences produced by highly parallel pyrosequencers. Nucleic Acids Res 36: e120.
64. Liu Z, Lozupone C, Hamady M, Bushman FD, Knight R (2007) Short pyrosequencing reads suffice for accurate microbial community analysis. Nucleic Acids Res 35: e120.

65. Lozupone C, Hamady M, Knight R (2006) UniFrac— an online tool for comparing microbial community diversity in a phylogenetic context. BMC Bioinformatics 7: 371.
66. Lozupone C, Knight R (2005) UniFrac: a new phylogenetic method for comparing microbial communities. Appl Environ Microbiol 71: 8228-8235.
67. Lozupone CA, Hamady M, Kelley ST, Knight R (2007) Quantitative and qualitative beta diversity measures lead to different insights into factors that structure microbial communities. Appl Environ Microbiol 73: 1576-1585.
68. Lozupone CA, Knight R (2008) Species divergence and the measurement of microbial diversity. FEMS Microbiol Rev 32: 557-578.
69. Ludwig W, Strunk O, Westram R, Richter L, Meier H, et al. (2004) ARB: a software environment for sequence data. Nucleic Acids Res 32: 1363-1371.
70. Magurran AE (2005) Biological diversity. Curr Biol 15: R116-R118.
71. Mardis ER (2008) The impact of next-generation sequencing technology on genetics. Trends Genet 24: 133- 141.
72. Margulies M, Egholm M, Altman WE, Attiya S, Bader JS, et al. (2005) Genome sequencing in microfabricated high-density picolitre reactors. Nature 437: 376-380.
73. Martin AP (2002) Phylogenetic approaches for describing and comparing the diversity of microbial communities. Appl Environ Microbiol 68: 3673-3682.
74. McKenna P, Hoffmann C, Minkah N, Aye PP, Lackner A, et al. (2008) The macaque gut microbiome in health, lentiviral infection, and chronic enterocolitis. PLoS Pathog 4: e20.
75. Melamede RJ, Wallace SS (1985) A possible secondary role for thymine-containing DNA precursors. Basic Life Sci 31: 67-102.
76. Miller SR, Augustine S, Olson TL, Blankenship RE, Selker J, et al. (2005) Discovery of a free-living chlorophyll dproducing cyanobacterium with a hybrid proteobacterial/ cyanobacterial small-subunit rRNA gene. Proc Natl Acad Sci USA 102: 850-855.
77. Oakley BB, Fiedler TL, Marrazzo JM, Fredricks DN (2008)

Diversity of human vaginal bacterial communities and associations with clinically defined bacterial vaginosis. Appl Environ Microbiol 74: 4898-4909.

78. Olsen GJ, Lane DJ, Giovannoni SJ, Pace NR, Stahl DA (1986) Microbial ecology and evolution: a ribosomal RNA approach. Annu Rev Microbiol 40: 337-365.

79. Pace NR (1997) A molecular view of microbial diversity and the biosphere. Science 276: 734-740.

80. Palmer C, Bik EM, Digiulio DB, Relman DA, Brown PO (2007) Development of the Human Infant Intestinal Microbiota. PLoS Biol 5: e177.

81. Pettersson E, Lundeberg J, Ahmadian A (2008) Generations of sequencing technologies. Genomics.

82. Pontes DS, Lima BCI, Chartone SE, maral Nascimento AM (2007) Molecular approaches: advantages and artifacts in assessing bacterial diversity. J Ind Microbiol Biotechnol 34: 463-473.

83. Pruesse E, Quast C, Knittel K, Fuchs BM, Ludwig W, et al. (2007) SILVA: a comprehensive online resource for quality checked and aligned ribosomal RNA sequence data compatible with ARB. Nucleic Acids Res 35: 7188- 7196.

84. Qiu X, Wu L, Huang H, McDonel PE, Palumbo AV, Tiedje JM, et al. (2001) Evaluation of PCR-generated chimeras, mutations, and heteroduplexes with 16S rRNA genebased cloning. Appl. Environ. Microbiol 67: 880-887.

85. Rainey FA, Ward RNL, Janssen PH, Hippe H, Stackebrandt E (1996) Clostridium paradoxum DSM 7308T contains multiple 16S rRNA genes with heterogeneous intervening sequences. Microbiology 142: 2087- 2095.

86. Riesenfeld CS, Schloss PD, Handelsman J (2004) Metagenomics: genomic analysis of microbial communities. Annu Rev Genet 38: 525-552.

87. Roesch LF, Fulthorpe RR, Riva A, Casella G, Hadwin AK, et al. (2007) Pyrosequencing enumerates and contrasts soil microbial diversity. ISME J 1: 283-290.

88. Ronaghi M (2001) Pyrosequencing sheds light on DNA sequencing.

Genome Res 11: 3-11.

89. Ronaghi M, Elahi E (2002) Pyrosequencing for microbial typing. J Chromatogr B Analyt Technol Biomed Life Sci 782: 67-72.

90. Ronaghi M, Karamohamed S, Pettersson B, Uhlen M, Nyren P (1996) Real-time DNA sequencing using detection of pyrophosphate release. Anal Biochem 242: 84-89.

91. Ronaghi M, Uhlen M, Nyren P (1998) A sequencing method based on real-time pyrophosphate. Science 281: 363-365.

92. Rossello MR, Amann R (2001) The species concept for prokaryotes. FEMS Microbiol Rev 25: 39-67.

93. Rothberg JM, Leamon JH (2008) The development and impact of 454 sequencing. Nat Biotechnol 26: 1117-1124.

94. Sanger F, Air GM, Barrell BG, Brown NL, Coulson AR, et al. (1977) Nucleotide sequence of bacteriophage phi X174 DNA. Nature 265: 687-695.

95. Schena M, Shalon D, Heller R, Chai A, Brown PO, et al. (1996) Parallel human genome analysis: microarraybased expression monitoring of 1000 genes. Proc Natl Acad Sci USA 93: 10614-10619.

96. Schloss PD, Handelsman J (2005) Introducing DOTUR, a computer program for defining operational taxonomic units and estimating species richness. Appl Environ Microbiol 71: 1501-1506.

97. Schloss PD Handelsman J (2006a) Introducing SONS, a tool for operational taxonomic unit-based comparisons of microbial community memberships and structures. Appl Environ Microbiol 72: 6773-6779.

98. Schloss PD, Handelsman J (2006b) Introducing TreeClimber, a test to compare microbial community structures. Appl Environ Microbiol 72: 2379-2384.

99. Schmeisser C, Steele H, Streit WR (2007) Metagenomics, biotechnology with non-culturable microbes. Appl Microbiol Biotechnol 75: 955-962.

100. Shendure J, Ji H (2008) Next-generation DNA sequencing. Nat Biotechnol 26: 1135-1145.

101. Simon GL, Gorbach SL (1986) The human intestinal microflora.

Dig Dis Sci 31: 147S-162S.

102. Sogin ML, Morrison HG, Huber JA, Mark WD, Huse SM, et al. (2006) Microbial diversity in the deep sea and the underexplored “rare biosphere”. Proc Natl Acad Sci USA 103: 12115-12120.

103. Sorek R,Zhu Y, Creevey CJ, Francino MP, Bork P, et al. (2007) Genome-wide experimental determination of barriers to horizontal gene transfer. Science 318: 1449-1452.

104. Spear GT, Sikaroodi M, Zariffard MR, Landay AL, French AL, et al. (2008) Comparison of the diversity of the vaginal microbiota in HIV-infected and HIVuninfected women with or without bacterial vaginosis. J Infect Dis 198: 1131-1140.

105. Spiegel CA (1991) Bacterial vaginosis. Clin Microbiol Rev 4: 485-502.

106. Spiegelman D, Whissell G, Greer CW (2005) A survey of the methods for the characterization of microbial consortia and communities. Can J Microbiol 51: 355-386.

107. Stahl DA, Lane DJ, Olsen GJ, Pace NR (1985) Characterization of a Yellowstone hot spring microbial community by 5S rRNA sequences. Appl Environ Microbiol 49: 1379-1384.

108. Stewart GC, Wilson FE, Bott KF (1982) Detailed physical mapping of the ribosomal RNA genes of Bacillus subtilis. Gene 19: 153-162.

109. Sundquist A, Bigdeli S, Jalili R, Druzin ML, Waller S, et al. (2007) Bacterial flora-typing with targeted, chipbased Pyrosequencing. BMC Microbiol 7: 108.

110. Suzuki MT, Giovannoni SJ (1996) Bias caused by template annealing in the amplification of mixtures of 16S rRNA genes by PCR. Appl Environ Microbiol 62: 625-630.

111. Tarnberg M, Jakobsson T, Jonasson J, Forsum U (2002) Identification of randomly selected colonies of lactobacilli from normal vaginal fluid by pyrosequencing of the 16S rDNA variable V1 and V3 regions. APMIS 110: 802-810.

112. Tringe SG, Hugenholtz P (2008) A renaissance for the pioneering 16S rRNA gene. Curr Opin Microbiol 11: 442-446.

113. Tringe SG, von MC, Kobayashi A, Salamov AA, Chen K, et al.

(2005) Comparative metagenomics of microbial communities. Science 308: 554-557.

114. Turnbaugh PJ, Hamady M, Yatsunenko T, Cantarel BL, Duncan A et al. (2008) A core gut microbiome in obese and lean twins. Nature.

115. Turnbaugh PJ, Ley RE, Mahowald MA, Magrini V, Mardis E.R, Gordon JI (2006) An obesity-associated gutmicrobiome with increased capacity for energy harvest. Nature 444: 1027-1031.

116. Turova TP, Kuznetsov BB, Novikova EV, Poltaraus AB, Nazina TN (2001) [Heterogeneity of nucleotide sequences of 16S ribosomal RNA genes from the Desulfotomaculum kuznetsovii type strain]. Mikrobiologiia 70: 788-795.

117. Vandamme P, Pot B, Gillis M, de VP, Kersters K, Swings J (1996) Polyphasic taxonomy, a consensus approach to bacterial systematics. Microbiol Rev 60: 407- 438.

118. Venter JC, Remington K, Heidelberg JF, Halpern AL, Rusch D, et al. (2004) Environmental genome shotgun sequencing of the Sargasso Sea. Science 304: 66-74.

119. von WF, Gobel UB, Stackebrandt E (1997) Determination of microbial diversity in environmental samples: pitfalls of PCR-based rRNA analysis. FEMS Microbiol Rev 21: 213-229.

120. Wang Q, Garrity GM, Tiedje JM, Cole JR (2007) Naive Bayesian classifier for rapid assignment of rRNA sequences into the new bacterial taxonomy. Appl Environ Microbiol 73: 5261-5267.

121. Wegley L, Edwards R, Rodriguez BB, Liu H, Rohwer F (2007) Metagenomic analysis of the microbial community associated with the coral Porites astreoides. Environ Microbiol 9: 2707-2719.

122. Whitman WB, Coleman DC, Wiebe WJ (1998) Prokaryotes: the unseen majority. Proc Natl Acad Sci USA 95: 6578-6583.

123. Wilson KH, Wilson WJ, Radosevich JL, DeSantis TZ, Viswanathan VS, et al. (2002) High-density microarray of small-subunit ribosomal DNA probes. Appl Environ Microbiol 68: 2535-2541.

124. Woese CR (1987) Bacterial evolution. Microbiol Rev 51: 221-271.

125. Woo PC, Ng KH, Lau SK, Yip KT, Fung AM, et al. (2003) Usefulness of the MicroSeq 500 16S ribosomal DNA-based

bacterial identification system for identification of clinically significant bacterial isolates with ambiguous biochemical profiles. J Clin Microbiol 41: 1996- 2001.

126. Woodmansey EJ (2007) Intestinal bacteria and ageing. J Appl Microbiol 102: 1178-1186.

127. Yergeau E, Schoondermark SSA, Brodie EL, Dejean S, DeSantis TZ, et al. (2008) Environmental microarray analyses of Antarctic soil microbial communities. ISME J.

128. Zhou X, Bent SJ, Schneider MG, Davis CC, Islam MR, et al. (2004) Characterization of vaginal microbial communities in adult healthy women using cultivationindependent methods. Microbiology 150: 2565-2573.

129. Zhou X, Brown CJ, Abdo Z, Davis CC, Hansmann MA, et al. (2007) Differences in the composition of vaginal microbial communities found in healthy Caucasian and black women. ISME J 1: 121-133.

130. Zoetendal EG, Rajilic SM, de Vos WM (2008) Highthroughput diversity and functionality analysis of the gastrointestinal tract microbiota. Gut 57: 1605-1615.

CHAPTER 10

DETECTION OF THREE CLOSELY LOCATED SINGLE NUCLEOTIDE POLYMORPHISMS IN THE EAAT2 PROMOTER: COMPARISON OF SINGLE-STRAND CONFORMATIONAL POLYMORPHISM (SSCP), PYROSEQUENCING AND SANGER SEQUENCING

Shavanthi Rajatileka[1], Karen Luyt[2,4], Maggie Williams[3], David Harding[4], David Odd[2,5], Elek Molnár[6] and Anikó Váradi[1]

[1]Centre for Research in Biosciences, Department of Biological, Biomedical and Analytical Sciences, Faculty of Health and Applied Sciences, University of the West of England, Bristol BS16 1QY, UK.

[2]Neonatal Neuroscience, School of Clinical Sciences, University of Bristol, St Michael's Hospital, Southwell Street, Bristol BS2 8EG, UK.

[3]Bristol Genetics Laboratory, Pathology Sciences, Blood Sciences and Bristol Genetics, Southmead Hospital, Bristol BS10 5NB, UK.

[4]Regional Neonatal Intensive Care Unit, St Michael's Hospital, University Hospitals NHS Trust, Bristol BS2 8EG, UK.

[5]Neonatal Intensive Care Unit, Southmead Hospital, North Bristol NHS Trust, Bristol BS10 5NB, UK.

[6]School of Physiology and Pharmacology University of Bristol, Medical Sciences Building, University Walk, Bristol BS8 1TD, UK.

ABSTRACT

Background

Single-strand conformational polymorphism (SSCP) is still a frequently used genotyping method across different fields for the detection of single nucleotide polymorphisms (SNPs) due to its simplicity, requirement for basic equipment accessible in most laboratories and low cost. This technique was previously used to detect rs4354668:A > C (g.-181A > C) SNP in the promoter of astroglial glutamate transporter (EAAT2) and the same approach was initially used here to investigate this promoter region in a cohort of newborns.

Results

Unexpectedly, four distinct DNA migration patterns were identified by SSCP. Sanger sequencing revealed two additional SNPs: g.-200C > A and g.-168C > T giving a rise to a total of ten EAAT2 promoter variants. SSCP failed to distinguish these variants reliably and thus pyrosequencing assays were developed. g.-168C > T was found in heterozygous form in one infant only with minor allele frequency (MAF) of 0.0023. In contrast, g.-200C > A and -181A > C were more common (with MAF of 0.46 and 0.49, respectively) and showed string evidence of linkage disequilibrium (LD). In a systematic comparison, 16% of samples were miss-classified by SSCP with 25-31% errors in the identification of the wild-type and homozygote mutant genotypes compared to pyrosequencing or Sanger sequencing. In contrast, SSCP and pyrosequencing of an unrelated single SNP (rs1835740:C > T), showed 94% concordance.

Conclusion

Our data suggest that SSCP cannot always detect reliably several closely located SNPs. Furthermore, caution is needed in the interpretation of the association studies linking only one of the co-inherited SNPs in the EAAT2 promoter to human diseases.

Keywords: EAAT2 promoter, Single nucleotide polymorphism, Genotyping, Pyrosequencing, SSCP, Premature newborns, Dried blood spots, Glutamate regulation

BACKGROUND

Genetic analysis is one of the fastest-growing areas of clinical diagnostics. Changes to a single nucleotide, known as single nucleotide polymorphism (SNP) is one of the major types of variants identified in the human genome. On average, in the human genome SNPs are distributed at 1 SNP per 1000 base pairs [1,2]. Some of these inherited SNPs play an important role in human diseases, while others are less relevant clinically and are phenotypically silent.

PCR amplification followed by Sanger DNA sequencing is one of the most commonly used methods of identifying SNPs in a sample cohort [3,4]. However, the cost per sample is still relatively high [5] and typically the sequencing run length is ~3 hours (based on genotyping ~700 bp amplicon using capillary array electrophoresis technology) [6]. Due to these drawbacks single-strand conformation polymorphism (SSCP) is still very frequently used across many different fields for SNP detection [7-15]. SSCP is a rapid, reproducible and quite simple method that does not require specialised expensive equipment or reagents. The SSCP process involves PCR amplification of the target fragment, denaturation of the double-stranded PCR product with heat and formamide and electrophoresis on a non-denaturing polyacrylamide gel. During electrophoresis the single-stranded DNA (ssDNA) fragments fold into three-dimensional shape depending on their primary sequence [7]. DNA fragments can then be genotyped as a result of their different migration patterns and then confirmed by Sanger sequencing. SSCP sensitivity varies considerably from 70% to 95% [16-19]. The disadvantages of this technique are that it is relatively labour intensive, low throughput and requires Sanger sequencing of a representative sample cohort to confirm the nucleotide sequence.

Pyrosequencing [20], a non-gel based, real-time, DNA sequencing-by-synthesis technique that is based on the luminometric detection of released pyrophosphate (PPi) during nucleotide incorporation, has also been used extensively for sample genotyping [21-26]. Pyrosequencing relies on a cascade of enzymatic reactions that yields detectable light, which is proportional to the incorporated nucleotides. The resulting pyrograms produce peak patterns in short stretches of the DNA sequence analysed, which vary between genotypes, and can distinguish between the different alleles at a named position. A large number of samples can therefore be analysed in a cost and time effective manner.

In this study, we investigated a previously identified SNP (rs4354668:A>C; [11]) in the promoter of the astroglial glutamate transporter *EAAT2* (SLC1A2) at position -181bp (g.-181A>C) in genomic DNA of newborn infants. The rational for looking at this particular SNP was that previous studies using SSCP found association of this SNP with increased extracellular glutamate levels and neurodegeneration in adult stroke patients [11]; with a higher risk of relapsing multiple sclerosis [27] and the progression of migraines into chronic daily headaches [28]. Unexpectedly, we identified two additional SNPs in the *EAAT2*promoter; g.-200C>A and g.-168C>T. The g.-168C>T SNP was only found in one individual in a heterozygous form in the entire cohort. In contrast, g.-200C>A and g.-181A>C sequence variants were much more common and they were in Linkage Disequilibrium (LD). SSCP was not discriminatory enough to clearly show differences between the various genotypes and 31% of homozygote mutants (mutant/mutant; MT/MT) and 25% wild-type (WT/WT) genotypes were identified incorrectly using this technique when compared to sequencing data. In contrast, pyrosequencing detected all naturally occurring variants in the highly GC-rich region and showed 100% concordance with Sanger sequencing suggesting that it can be used successfully to detect closely positioned and linked SNPs. Our data also indicate that the interpretation of the studies [11] attributing a causal link between g.-181A>C and adult neurological diseases is incomplete as the SNP was potentially misclassified and the LD with another SNPs not considered.

METHODS

Sample collection and processing

Newborn dried blood spots (DBS) were collected from predominantly Caucasian infants (91.6% white, 8.4% non-white) born in the greater Bristol area (UK) participating in an association study to investigate the genetic background of newborn infants to white matter brain injury. The study received ethical approval in April 2008 from the National Research Ethics Service, UK (REC reference number 10/H0106/10 [29]). Samples, collected from 239 infants within the past 3-22 years, were used in the study. All blood spot screening cards were stored in the biobank in boxes at room temperature. Whole blood samples were collected from nine healthy adult volunteers to optimise protocols used in the study. Genomic DNA was isolated and quantified as we described previously [29].

PCR amplification of EAAT2 promoter for SSCP analysis

Previously described primers EAAT2F and EAAT2R were used to amplify the EAAT2 promoter fragment (GeneBank accession AF510107.1; Figure 1 and Table 1[11]). All PCR reactions were carried out for 35 cycles in a total volume of 25 μl, containing 1× high fidelity reaction buffer - (500 mM KCl, 100 mM Tris-HCl, pH 8.3), 1 mM of $MgCl_2$, 200 μM of each dNTP, 100 pmol of each oligonucleotide primer, 1 unit of high fidelity Taq Polymerase (FastStart High Fidelity Taq Polymerase, Roche Diagnostics Limited, West Sussex, UK) and 2 μl (~1-30 ng) of gDNA. Additionally, a final concentration of 1× GC-rich solution (Roche Diagnostics Limited, West Sussex, UK) was added to each reaction. Reaction parameters were 95°C for 5 min followed by 35 cycles of 95°C for 30 s, 60°C for 45 s and 72°C for 1 min. A final extension at 72°C was carried out for 10 min.

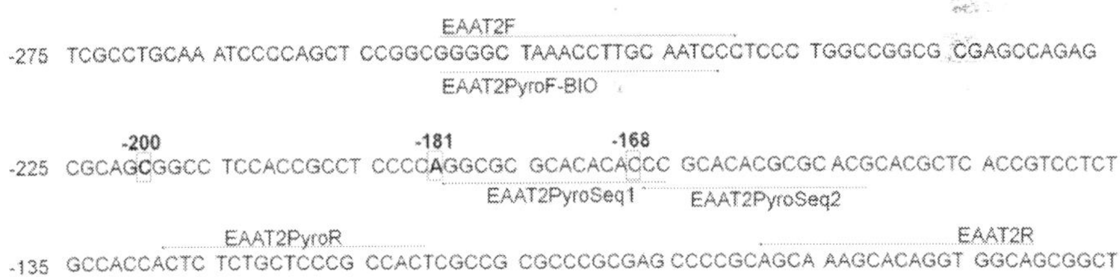

Figure 1: Promoter sequence of the human *EAAT2* (Accession AF510107.1). The primers and the positions of the three SNPs at -200bp (g.-200C > A), -181bp (g.-181A > C) and -168bp (g.-168C > T) are indicated. Numbering is relative to the transcription start site. Primers EAAT2F and EAAT2R were used for standard PCR and Sanger sequencing while EAAT2PyroF-BIO and EAAT2Pyro-R were used to generate biotinylated PCR products and EAAT2PyroSeq1 and EAAT2PyroSeq2 for pyrosequencing (see also Table 1).

Table 1: Pyrosequencing primers and conditions used in the study

Oligonucleotide	Sequence 5′-3′	Product (bp)	Annealing T (°C)	Annealing T (°C)
EAAT2F	GGGGCTAAACCTTGCAATCC	180	65	None
EAAT2R	CTGCCACCTGTGCTTTGC			
EAAT2PyroF-BIO	GGGGCTAAACCTTGCAATC	166	60	5′ Biotin
EAAT2PyroR	GAGTGGCGGGAGCAGAGA			None
EAAT2PyroSeq1	GGGTGTGTGCGCGCC		N/A	None
EAAT2PyroSeq2	CCGCACACGCGCACG		N/A	None
Target sequence for pyrosequencing (1)		**T/G**GGGGAGGCGGTGGAGGCC**G/T**CTG		
Nucleotide dispensation order (1)		CGTGCAGCGTGAGCGTGC		
Target sequence for pyrosequencing (2)		**G/A**TGTGTGCGCGCC		
Nucleotide dispensation order (2)		CAGTGTGT		

Primer pair EAAT2PyroF-BIO/EAAT2PyroR were used to generate biotinylated PCR products flanking SNPs g.-200C > A; g.-181A > C and g.-168C > T. Primers EAAT2PyroSeq1 (to detect g.-200C > A;-181A > C) and EAAT2PyroSeq2 (to detect g.-168C > T) were used for pyrosequencing. In the dispensation order the nucleotides used as negative controls are underlined. The nucleotide change in the Target sequence for pyrosequencing is indicated in bold.

SSCP analysis

SSCP was performed as previously described [8]. PCR samples were resolved on 0.5× acrylamide gels containing 12.5 ml MDE® (Mutation Detection Enhancement) gel solution (Lonza Group Ltd., Basel, Switzerland), 3 ml of 10× TBE (Tris/Borate/EDTA, pH 8.3) buffer, 34.28 ml deionised water, 20 μl tetramethylethylenediamine (TEMED; Sigma-Aldrich, St Louis, Missouri, UK) and 200 μl of freshly prepared 10% ammonium persulfate (APS; Sigma-Aldrich, St Louis, Missouri, UK). PCR samples were prepared for electrophoresis as follows; 3 μl of PCR product was mixed with 7 μl of denaturing loading buffer (95% formamide, 0.025% bromophenol blue, 0.025% xylene cyanol and 20 mM EDTA) (all reagents from Sigma-Aldrich, St Louis, Missouri, UK). The mixture was heated to 95°C for 5 min, rapidly cooled on ice and then 10 μl was loaded and run for 30 min at 300 V. The voltage was then reduced to 150 V and the DNA strands separated for 14 h at room temperature (~20°C). The gel was washed twice in distilled water for 10 s and then incubated in 0.5% glacial acetic acid (Fisher Scientific, Loughborough, UK) and 10% molecular grade ethanol (Sigma-Aldrich, St Louis, Missouri, UK). The gel was then incubated in 0.1% silver nitrate (Sigma-Aldrich, St Louis, Missouri, UK) solution for 20 min and rinsed with distilled water twice. The gel was then washed with developing solution, 1.5% NaOH (Fisher Scientific, Loughborough, UK) and 0.15% molecular grade formaldehyde (Sigma-Aldrich, St Louis, Missouri, UK) for 20 min. The gel was fixed in 0.75% sodium carbonate (Fisher

Scientific, Loughborough, UK) solution for 10 min. The DNA bands were visualized on a light box and the samples were scored.

Generation of biotinylated PCR products for pyrosequencing

Two sequence-specific primers (EAAT2PyroF-BIO and EAAT2PyroR; Figure 1, Table 1) were designed to flank all SNPs in the *EAAT2* promoter using the software provided by Qiagen Pyrosequencing, with the forward primer biotinylated. PCR reactions contained 1× PCR buffer (500 mM KCl, 100 mM Tris-HCl, pH 8.3), 1.5 mM $MgCl_2$, 200 μM of each dNTP, 100 pmol of each oligonucleotide and 1 unit of high fidelity Taq polymerase (FastStart High Fidelity Taq Polymerase, Roche Diagnostics Limited, West Sussex, UK) per reaction. Two microlitres of genomic DNA (containing 4-6 ng DNA) was used per reaction. Amplification was performed with the following conditions: 95°C for 5 min; 50 cycles of 94°C for 30 s, 60°C for 30 s and elongation at 72°C for 30 s; followed by the final extension for 10 min at 72°C. Pyrosequencing and Sanger sequencing were carried out as we described previously [29]. The target sequence for analysis and the nucleotide dispensation order for the pyrosequencing assay are shown in Table 1. Purified PCR products were Sanger sequenced using primer EAAT2R (Table 1).

RESULTS

Analysis of the EAAT2 promoter using SSCP

A SNP was detected in the *EAAT2* promoter at -181bp by SSCP [11]. Since we were interested in this promoter region and already had considerable expertise in this method [8], we used SSCP for our initial experiments. Although it is not possible to predict the three-dimensional structure from the primary sequence of the ssDNA [19], it is expected that the wild-type (WT/WT), mutant (MT/MT) and heterozygote (WT/MT) would have a unique electrophoretic mobility. Indeed, our SSCP result showed the expected three distinct patterns (Figure 2, Lanes 1-3). However, one sample (Figure 2, Lane 4) showed some unexpected extra bands. Sanger sequencing of samples scored based on their migration pattern as wild type (n=4, Lane 1), heterozygotes (n=2, Lane 2), homozygote mutants (n=2, Lane 3) and the sample with an unusual DNA migration (n=1, Lane 4) revealed a previously unpublished polymorphism C to A transition at -200bp (g.-200C>A), 19 bp upstream from the A to C transition observed at -181 bp (g.-181A>C; [11]).

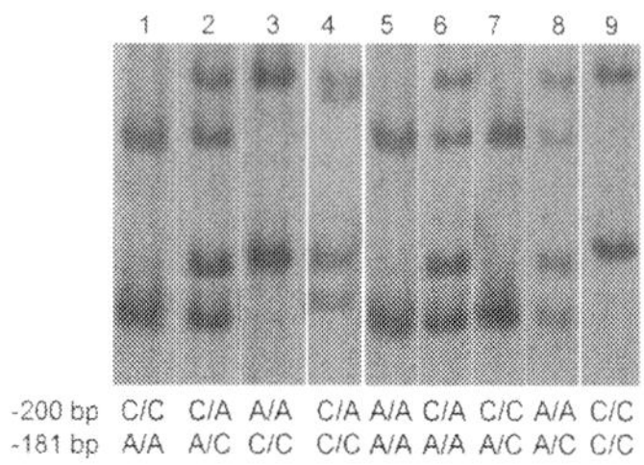

Figure 2: SSCP patterns of the *EAAT2* promoter genotypes. Following PCR amplification, all samples were run on the same SSCP gel and then visualised. The genotype of each sample determined by Sanger sequencing is shown at the bottom of each lane. Note that all these samples were wild type for g.-168C>T.

Sanger sequencing also revealed that the following additional genotypes exist (sequence is given in -200 bp and -181bp order): A/A and A/A (Figure 2, Line 5); C/A and A/A (Figure 2, Line 6); C/C and A/C (Figure 2, Lane 7); A/A and A/C (Figure 2; Line 8); C/C and C/C (Figure 2, Line 9). These variants did not migrate differently compared to the three main types (Figure 2, Lines 1-3), even when the SSCP running conditions were further optimised suggesting that this technique is unsuitable for the detection of all nine possible *EAAT2* variants (Figure 2).

Optimization of pyrosequencing to detect all EAAT2 variants

The SSCP revealed that it was essential to get sequencing data for all samples for accurate genotyping. Thus, we used pyrosequencing, which is suitable for the amplification of this short region and provides exact sequence data for a large number of samples. Pyrosequencing was optimised and evaluated using genomic DNA prepared from blood from healthy adult volunteers. The initial assay was designed to use the forward strand but this approach was unsuccessful and the reading failed at the SNP g.-181A>C (Figure 1, Figure 3A and B left panel). Therefore, pyrosequencing was carried out on the reverse strand which generated clear pyrograms (Figure 3B right panel, Figure 4). Note that the sequence is given in reverse orientation.

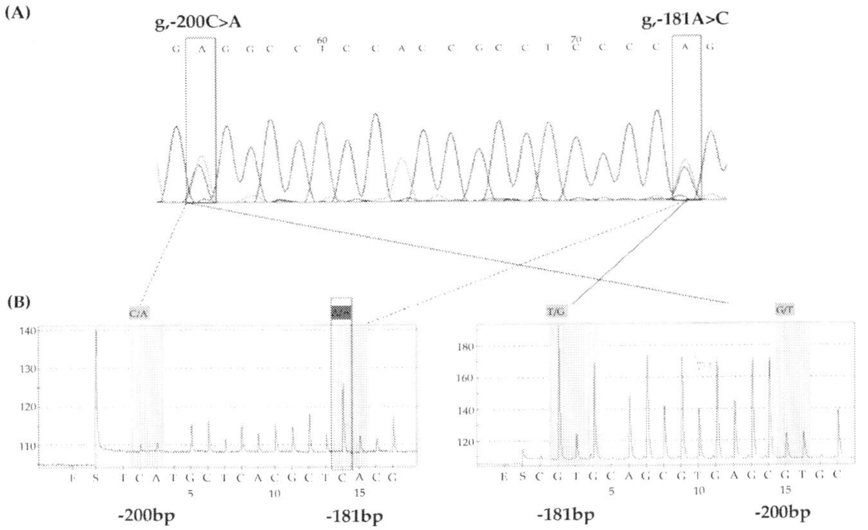

Figure 3: Pyrograms using forward and reverse strands for sequencing. (A). SNPs g.-200C > A;-181A > C are indicated in rectangles on the Sanger sequence traces. (B) Pyrograms of the same sample using forward (left panel) and reverse (right panel) primers. Arrows indicate the region sequenced by both methods.

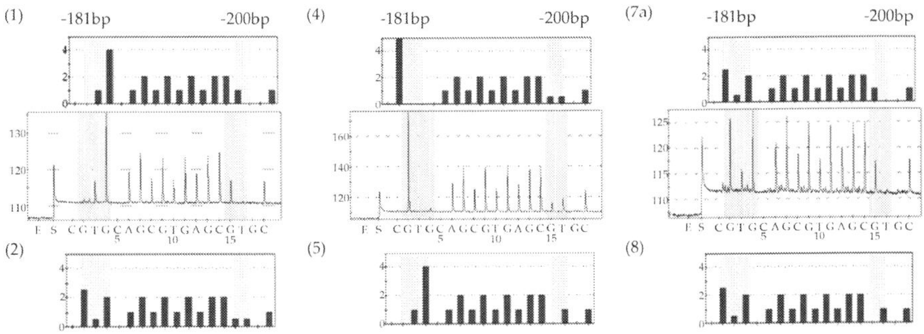

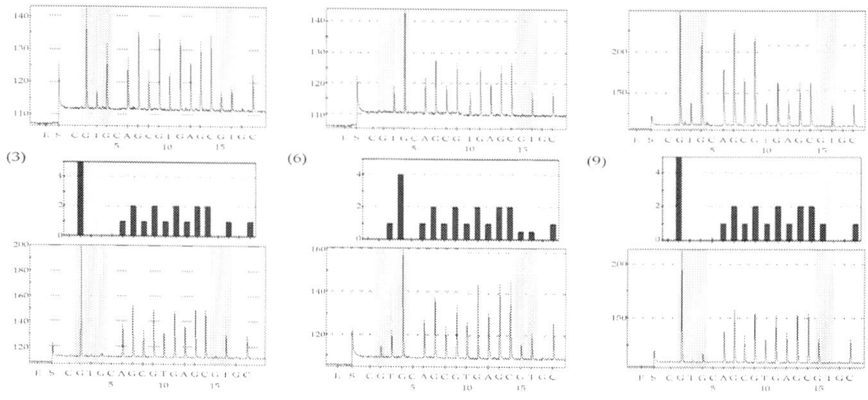

Figure 4: Predicted (top panels) and observed (bottom panels) pyrograms for *EAAT2* promoter SNPs. The position of the SNPs is highlighted in yellow boxes, the x-axis of each pyrogram indicates the order of reagent addition (E - enzyme, S -substrate and nucleotide A,G,T or C); the y-axis shows the light intensity generated. The numbering of pyrograms corresponds to the haplotype numbers in Table 2. Note that all these samples were wild type for g.-168C > T.

Polymorphism analysis of the EAAT2 promoter using pyrosequencing

Successful amplification was obtained in 209 samples (87.5% success rate). Failure of the remaining samples was likely due to low quality genomic DNA. Some of the samples were 22 years old and showed DNA degradation [29]. Overall in 89% of the samples the polymorphisms g.-200C > A;-181A > C were inherited together (Table 2). While the SSCP data indicates that the genotype distribution of these SNPs is in Hardy Weinberg Equilibrium (HW), the pyrosequencing results suggest the opposite (Table 3). Measures of LD (D' and r^2), the non-random association between alleles of different loci, are consistent with the SNPs being linked (Table 3). The analysis and interpretation of LD is difficult due to the lack of HW and the presence of only one mutation at the -168 loci. Haplotype predictions are also shown in Table 4. A 100% concordance was observed when compared with Sanger sequencing (n = 51 samples were sequenced with both methods).

Table 2: Distribution of genotypes in the sample cohort

Genotype	−200C > A	−181A > C	−168C > T	Number & proportion
1	C/C	A/A	C/C	42 (20%)
2	C/A	A/C	C/C	110 (53%)
3	A/A	C/C	C/C	34 (16%)
4	C/A	C/C	C/C	7 (3%)
5	A/A	A/A	C/C	4 (2%)
6	C/A	A/A	C/C	0
7a	C/C	A/C	C/C	9 (4.5%)
7b	C/C	A/C	C/T	1 (0.5%)
8	A/A	A/C	C/C	2 (1%)
9	C/C	C/C	C/C	0
		n = 209		
WT	C = 0.54	A = 0.51	C = 0.997	Allele frequency
MT	A = 0.46	C = 0.49	T = 0.002	

Genotypes were identified by pyrosequencing (n = 209) and confirmed by Sanger sequencing (n = 51). WT – wild type; MT – mutant.

Table 3: Hardy Weinberg equilibrium and LD variance for the three *EAAT2* SNPs using pyrosequencing or SSCP

	Pyrosequencing	SSCP
	Hardy-Weinberg	
for -168	p > 0.99	N/A
for -181	p = 0.0188	p = 0.4063
for -200	p = 0.0256	p = 0.3325
	Lewontin's D′	
−168 and -181	0.49	N/A
−168 and -200	1	N/A
−181 and -200	0.94	1
	r^2	
−168 and -181	0	N/A
−168 and -200	0	N/A
−181 and -200	0.79	0.99

Lewontin's D′ and r^2 both give ordinal measures of Linkage Disequilibrium (LD). Please note only one mutation was found at -168 making interpretation difficult for these associations.

Table 4: Predicted haplotype frequencies in the cohort using pyrosequencing or SSCP

Pyrosequencing										
SNP	Haplotype								C allele	Total alleles
−168 C > T	T	T	T	T	C	C	C	C	417	418
−181 A > C	A	A	C	C	A	A	C	C	204	418
−200 C > A	A	C	A	C	A	C	A	C	225	418
%	0.0	0.0	0.0	0.2	1.5	49.8	44.6	3.9		

SSCP						
SNP	Haplotype				C allele	Total alleles
−181 A > C	A	A	C	C	208	414
−200 C > A	A	C	A	C	205	414
%	0.2	49.5	50.2	0.0		

(Total number of 'C' alleles is indicated).

Comparison of sample genotyping using pyrosequencing and SSCP

All nine sequence combinations have been successfully amplified and pyrosequenced (Figure 4; Note that genotypes six and nine were only found in the adult control samples hence they are not presented in Tables 2, ,5 and.6). 239 samples from newborn infants were initially used and 209 could be classified for SSCP and pyrosequencing. Because different samples failed to produce clear PCR products for SSCP and pyrosequencing, a total of 183 samples generated result with both genotyping methods. With SSCP a total of 29 samples (16%) were incorrectly genotyped (Table 5). While 51 samples were classified as homozygote wild type using SSCP, pyrosequencing revealed that 25% of these samples do not belong to this group (Table 5). There was surprisingly little error in the identification of the heterozygotes with SSCP and 96% of the samples were correctly genotyped. In contrast, 12 homozygote mutants (31%) were incorrectly identified (Table 5). We also genotyped a small number of samples (n = 15) that failed to produce a clear PCR product with the EAAT2F and EAAT2R primers (hence could not be used for SSCP, Figure 1) but resulted in clear pyrograms with the EAAT2PyroSeq1 primer. A second SSCP was carried out with EAAT2F and EAAT2PyroR primers (Table 6) and found that with these primers similar proportion (20% *versus* 16%) of samples were misclassified as with the EAAT2F and EAAT2R primers.To compare the concordance between SSCP and pyrosequencing for a single SNP, SNP rs1835740 was analysed in the same 239 samples. Three distinct

SSCP patterns were observed for the different genotypes (Figure 5A) which were confirmed by a random Sanger sequencing (Figure 5B) and pyrosequencing (Figure 5C) of the whole cohort. The concordance rate between SSCP and pyrosequencing was 94% for this SNP.

Table 5: Comparison of genotypes identified by SSCP and pyrosequencing

SSCP	Pyprosequencing								
Genotype	1	2	3	4	5	6	7	8	9
1. WT/WT (n = 51)	**38 (74.5%)**	1 (1.9%)	0.0	0.0	4 (7.8%)	0.0	8 (15.6%)	0.0	0.0
2. WT/MT (n = 92)	0.0	**88 (95.6%)**	2 (2.2%)	0.0	0.0	0.0	0.0	2 (2.2%)	0.0
3. MT/MT (n = 39)	0.0	6 (15.3%)	**27 (69.2%)**	6 (15.4%)	0.0	0.0	0.0	0.0	0.0
4. (n = 1)	0.0	0.0	0.0	**1.0**	0.0	0.0	0.0	0.0	0.0
n = 183				n = 183					

For the SSCP EAAT2F and EAAT2R primers were used and pyrosequencing was done with EAAT2PyroSeq1 primer (Figure 1 and Table 1). The genotypes that were correctly identified by both methods are indicated in bold. The genotype numbers indicated under pyrosequencing corresponds to those used in Table 2.

Table 6: Comparison of genotypes identified by SSCP and pyrosequencing

SSCP	PYROSEQUENCING								
Genotype	1	2	3	4	5	6	7	8	9
1. WT/WT (n = 3)	**2 (66.7%)**	0.0	0.0	0.0	0.0	0.0	1 (33.3%)	0.0	0.0
2. WT/MT (n = 11)	0.0	**9 (81.8%)**	1 (9.1%)	0.0	0.0	0.0	1 (9.1%)	0.0	0.0
3. MT/MT (n = 1)	0.0	0.0	**1 (100%)**	0.0	0.0	0.0	0.0	0.0	0.0
n = 15				n = 15					

For SSCP EAAT2F and EAAT2Pyro primers were used and pyrosequencing was done with EAAT2PyroSeq1 primer. The genotypes that were correctly identified by both methods are indicated in bold.

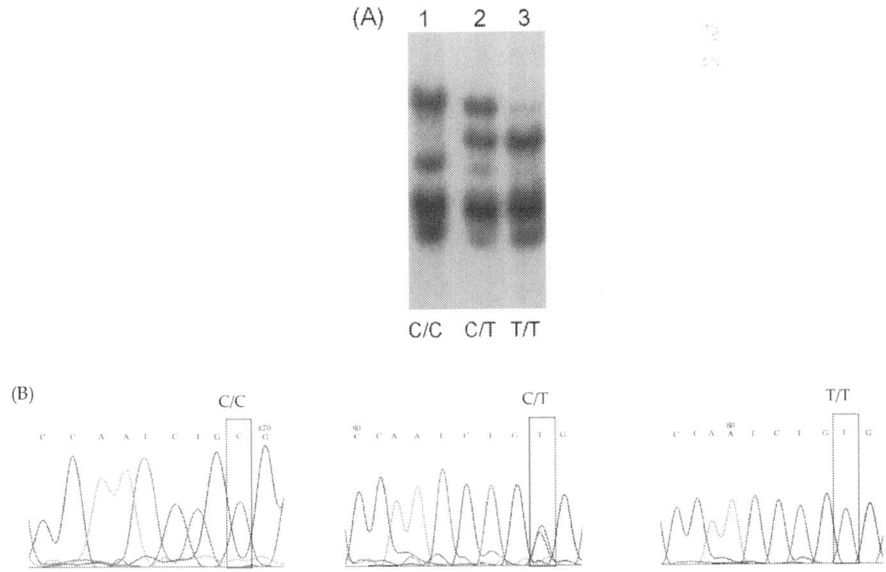

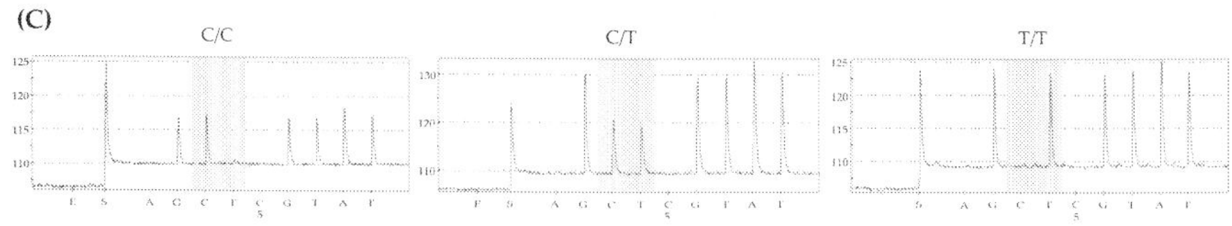

Figure 5: Detection of an unrelated SNP, rs1835740, by SSCP, Sanger sequencing and pyrosequencing. (A) Genotype of each sample determined by Sanger sequencing is shown at the bottom of each lane. (B) SNP is indicated in rectangles on the Sanger sequence. **(C)** The position of the SNP on the pyrogram is highlighted in yellow boxes.

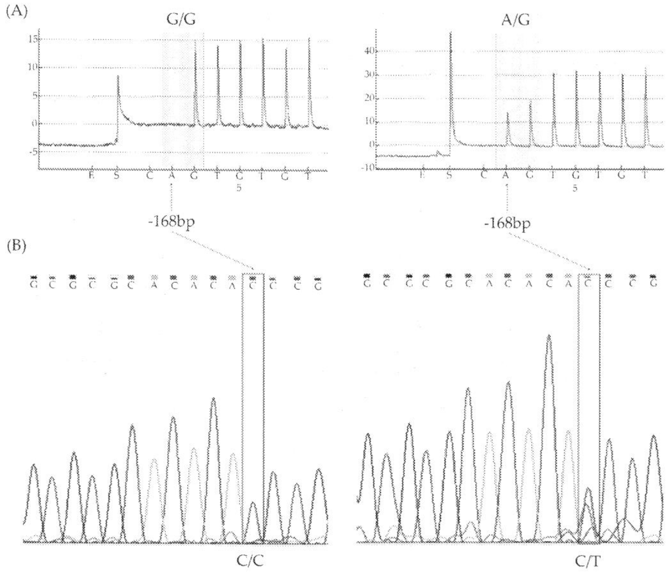

Figure 6: Detection of the SNP in -168bp (g.-168C > T) in the *EAAT2* promoter. Pyrogram (A) and Sanger sequencing (B) of the homozygote WT and heterzygote samples.

While our investigation was underway, a SNP g.-168C>T was entered into the Database of Single Nucleotide Polymorphisms (dbSNP), through the 1000 Genomes Project [30] and was given a reference number of (rs116392274:C>T; Human Build 137). This nucleotide change is located in the EAAT2PyroSeq1 primer sequence (Figure 1) and thus it could not be observed in the pyrograms. However, using Sanger sequencing 51 samples were sequenced with EAAT2R (Figure 1) and in all of these samples only the C allele was observed at position -168bp. Furthermore, a pyrosequencing assay was developed to detect this g.-168C > T specifically (Figure 6). Of the analysed samples, 213 were wild type (C/C) and one sample was a heterozygote (C/T) for this SNP. The MAF was 0.0023 in our cohort.

DISCUSSION

Identification of additional SNPs in the EAAT2 promoter

We identified a polymorphism at -200bp in the EEAT2 promoter, 19bp upstream of the previously reported and characterised polymorphism at -181bp (rs4354668:A>C or g.-181A>C) [11]. Our data indicates that these SNPs are in LD (Table 3). While our study was close to completion, the g.-200C > A was added to the NCBI SNP Database (1000 Genome Project, Human Build 137; rs111885243:C > A) confirming our sequencing and pyrosequencing data. The MAF in our predominantly Caucasian cohort for g.-200C>A and g.-181A>C is 0.46 and 0.49, respectively. The Global MAF available from the SNP Database are 0.39 and 0.41, respectively. More recently another SNP in the *EAAT2* promoter at position -168bp was added to the NCBI SNP Database (Human Build 137; rs116392274:C>T). In the 51 samples that we sequenced only the C allele was present. Furthermore, in the entire cohort (n=214) only one T allele was found in a heterozygous form (Table 2). To date, these newly identified SNPs (g.-200C>A and g.-168C>T) have not been investigated in association studies or cited in the literature.

SSCP is not sensitive enough to reliably distinguish between the various EAAT2 promoter genotypes

SSCP was used initially in this study because this method has previously been applied to genotype exactly the same region of the *EAAT2* promoter [11]. We used the same primers and PCR conditions as reported [11] but modified the SSCP running conditions that provided better separation of the DNA strands. Previously, approximately 2 h at a high voltage was used to resolve the amplicons. In contrast, in the current study the PCR products were resolved for 14 h at a relatively low voltage (150V) at a constant temperature (20°C). This allowed better separation and visualization of the ssDNA bands and lead to the identification of an additional genotype (Figure 2, Lane 4). Sequencing of several samples lead to the identification of g.-200C > A, which was not reported in a previous study of this region [11]. Our SSCP, pyrosequencing and Sanger sequencing highlighted that although four clear migration patterns can be seen (Figure 2, Lanes 1-4) several of the other variants (Figure 2, genotypes 5-9,) could not be identified by SSCP. Note that the reproducibility of SSCP was 100% for the samples that were used as controls (one sample from each of the three main genotypes were always run on each gel, in total n = 45 samples).

Studies using SSCP showed that the position of the substitution within a codon and the nucleotide itself can determine whether a SNP is detected [31]. A G to A or G to T nucleotide change at the second position of a codon caused a shift in ssDNA migration while failed to do so if it occurred in the first position [31]. In our case the SNP at -181bp is located on the second base, while the SNP at -200bp is on the first base of a codon. Furthermore, some nucleotide changes are detected at lower rates than others. For example, A to C transversions were detected at a higher rate (95%) compared to C to A transversions (82%) [31]. The SNP at -200bp is a C to A whilst the SNP at -181bp is an A to C transversion. It is also documented that some point mutations are not detected because of the nucleotide composition (e.g. A + T or G + C richness) of a DNA region being analysed [32]. Indeed, the *EAAT2* promoter is highly GC-rich (Figure 1). The amplicon used in both the previous study [11] and this study for SSCP analysis has a GC content of ~73%. Furthermore, some mutations may cause relatively small changes in electrophoretic mobility [33] and might remain undetected by SSCP [34-36]. These factors could explain that the SSCP patterns for the *EAAT2* promoter resemble that of a single SNP instead of multiple SNPs. However, the banding pattern does not fully correspond to the genotype of the SNP at -181bp. While genotypes 5 and 8 followed the -181bp SNP migration pattern, genotypes 6 and 7 resembled the migration of the -200bp SNP.

Based on the SSCP analysis, 25-31% of the WT/WT and MT/MT samples were mis-classified (Table 5 and .6). The previous study of the *EAAT2* promoter region [11] identified only three SSCP patterns in their cohort. However, considering the MAF of g.-200C > A and g.-181A > C in the population (0.46 and 0.49 in the current predominantly Caucasian cohort; 0.39 and 0.41 in the SNP Database), it is expected that some of the additional variants described here, should have been identified in the previous study (Table 4). Indeed, a similar allele frequency and LD levels are expected in Caucasian cohorts [37]. Furthermore, numerous subsequent studies [27,28,38,39] understandably continued with only investigating the association of this single SNP (rs4354668:A>C or g.-181A>C) with various diseases.

Many studies across different fields still use SSCP extensively as a genotyping method and about 1040 studies are listed on PubMed that used SSCP since 2010 to date. It is a simple, user-friendly, low cost method of SNP detection which does not require specialist equipment and can be adapted to a high-throughput format. It can work very effectively when a single SNP is investigated as we have demonstrated for an unrelated single SNP, rs1835740:C>T (Figure 5). However, our results also highlight that SSCP cannot always be used effectively when several SNPs are located in the target sequence. Although it is well recognised that a representative sample with distinct SSCP pattern needs to be sequenced to validate the method, it is also crucial that the entire sequence of the PCR product used for SSCP is scrutinised carefully. Generation of shorter PCR products for SSCP can sometimes help to uncover previously unnoticed variants [36,40]. If SSCP is used for genotyping (not for mutational screening) and all SNPs in the regions are known, covering some of them with primers may eliminate them from the SSCP pattern making the analysis of the remainders easier. The PCR products for the *EAAT2* promoter are already short, generating even shorter targets thus would not solve the problem seen with this particular target sequence but might offer solution for other troublesome targets.

Pyrosequencing as an alternative to detect closely positioned SNPs

In our study g.-200C>A;-181A>C could simultaneously be analysed by pyrosequencing (Figures 3, 4 and Table 2). The detection limit of this method is dependent on how well the dispensation profile can be set up. This in turn depends on the nucleotide change within the SNP and the nucleotides adjacent to the SNP(s) [41]. Indeed, the latter caused problems in the genotyping of g.-181A>C using a primer in 5′-3′ orientation (Figure 3B, left panel). A four C mononucleotide repeat precedes this

SNP and the non-linear light generation of the mononucleotide repeat made it impossible for the software to interpret the correct number of incorporated identical nucleotides [42,43] and as a consequence the assay failed at the g.-181A > C SNP (in 100% of the 96 samples analysed). This problem was overcome by re-designing the assay on the reverse strand and sequencing the nucleotide change prior to the C mononucleotide repeats (Figure 3B, right panel). Similarly, g.-168C > T was also sequenced on the reverse strand (Figure 1). Both pyrosequencing assays generated sequences immediately downstream of the primer (Figure 1, 3 and ,6), which cannot be achieved with Sanger sequencing that lays a reading gap of 20-30 bp from the sequencing primer [44]. Pyrosequencing can only analyse a few positions simultaneously [41], which was the main reason for developing two separate assays to detect g.-200C > A;-181A > C and g.-168C > T (Figure 1). This approach resulted in clear and distinguishable pyrograms for each genotype for each assay. g.-168C > T was found in a heterozygote form in one infant with no clinical evidence of white matter injury (Rajatileka et al. unpublished observation). The MAF of the g.-168C > T (0.0023 in our study and 0.017 on the SNP Database) is very low in the general population which makes it challenging to assess in association studies. SNP-SNP interactions have been suggested to have a great impact on unveiling the underlying mechanism of complex diseases [45]. Thus, future clinical investigations of the impact of g.-200C > A;-181A > C on the promoter function of *EAAT2* and their association with various diseases will need to be assessed simultaneously.

Currently, the detection of g.-200C > A;-181A > C cost £1.79 and £1.43 by pyrosequencing and SSCP, respectively. For pyrosequencing the cost includes all reagents and a charge for the use of the pyrosequencer. For SSCP the cost was calculated from the reagents and Sanger sequencing of 10% of the samples. Following PCR amplification, the pyrosequencing required 1 h preparation time and 21 min run time for the two SNPs for 96 samples. For a single SNP (such as rs1835740:C > T) the run time is usually ~10 min for 96 samples. In contrast, SSCP analysis of 100 samples requires 2-3 h post-PCR preparation time, 12-16 h gel electrophoresis and 0.45-1.5 h silver staining. In addition, at least 10% of the samples need to be prepared for Sanger sequencing. Whilst pyrosequencing provides a less labour intensive, low cost and high throughput platform to genotype samples, in laboratories with no access to this facility SSCP may be used reliably for genotyping if *(i)* all mutations in the region are known, *(ii)* the SSCP genotype readout is validated by another method, and *(iii)* in case of two SNPs in the region the indicative bands for both mutations are clearly and easily distinguishable.

CONCLUSION

Our data suggest that SSCP cannot always detect reliably several closely located SNPs. Furthermore, caution is needed in the interpretation of the association studies linking only one of the co-inherited SNPs in the EAAT2 promoter to human diseases.

AUTHORS' CONTRIBUTIONS

SR designed and carried out all experimental work and data analysis. DO carried out the statistical analysis. KL and DH arranged access to the adult and newborn clinical samples, and contributed to clinical study design. KL obtained research ethics, NHS R&D permissions, University of Bristol research sponsorship for use of human tissue and consenting processes. MW assisted with the pyrosequencing analysis. EM and AV advised on experimental design. SR and AV wrote the manuscript and all authors reviewed the manuscript prior to submission. All authors read and approved the final manuscript.

ACKNOWLEDGEMENTS

This project was funded by the University of the West of England, Bristol, UK (Grant awarded to AV). EM is supported by the Biotechnology and Biological Sciences Research Council, UK (grants BB/F011326/1 and BB/J015938/1). The blood spot retrieval was funded by the David Telling Charitable Trust. We would like to thank Dr Helena Kemp and the NHS Newborn Screening laboratory for assisting with retrieving samples from the repository.

REFERENCES

1. Landegren U, Nilsson M, Kwok PY. Reading bits of genetic information: methods for single-nucleotide polymorphism analysis. Genome Res. 1998;8:769–776.

2. Wang DG, Fan JB, Siao CJ, Berno A, Young P, Sapolsky R, Ghandour G, Perkins N, Winchester E, Spencer J, Kruglyak L, Stein L, Hsie L, Topaloglou T, Hubbell E, Robinson E, Mittmann M, Morris MS, Shen N, Kilburn D, Rioux J, Nusbaum C, Rozen S, Hudson TJ, Lipshutz R, Chee M, Lander ES. Large-scale identification, mapping, and genotyping of single-nucleotide polymorphisms in the human genome. Science. 1998;280:1077–1082.
3. Marzinotto S, Sessa F, Franzoni A, Anselmi A, Gastaldo LR, Mason S, Damante G, Beltrami CA, Mariuzzi L. KRAS codons 12 and 13 mutation analysis: a comparative study between direct sequencing and a new sensitive real-time PCR assay. Sequencing. 2011;2011:895709.
4. Angulo B, Conde E, Suárez-Gauthier A, Plaza C, Martínez R, Redondo P, Izquierdo E, Rubio-Viqueira B, Paz-Ares L, Hidalgo M, López-Ríos F. A comparison of *EGFR* mutation testing methods in lung carcinoma: direct sequencing. Real-time PCR and immunohistochemistry. PLoS ONE. 2012;7:e43842.
5. Kircher M, Kelso J. High-throughput DNA sequencing concepts and limitations. Bioessays. 2010;32:524–536.
6. Morozova O, Marra MA. Applications of next-generation sequencing technologies in functional genomics. Genomics. 2008;92:255–264.
7. Orita M, Iwahana H, Kanazawa H, Hayashi K, Sekiya T. Detection of polymorphisms of human DNA by gel electrophoresis as single-strand conformation polymorphisms. Proc Natl Acad Sci U S A. 1989;86:2766–2770.
8. Varadi A, Lebel L, Hashim Y, Mehta Z, Ashcroft SJ, Turner R. Sequence variants of the sarco(endo)plasmic reticulum $Ca^{(2+)}$-transport ATPase 3 gene (SERCA3) in Caucasian type II diabetic patients (UK Prospective Diabetes Study 48) Diabetologia. 1999;42:1240–1243.
9. Bettinaglio P, Galbusera A, Caprioli J, Orisio S, Perna A, Arnoldi F, Bucchioni S, Noris M. Single strand conformation polymorphism (SSCP) as a quick and reliable method to genotype M235T polymorphism of angiotensinogen gene. Clin Biochem. 2002;35:363–368.
10. Daly AK, King BP, Leathart JB. Genotyping for cytochrome P450

polymorphisms. Methods Mol Biol. 2006;320:193–207.

11. Mallolas J, Hurtado O, Castellanos M, Blanco M, Sobrino T, Serena J, Vivancos J, Castillo J, Lizasoain I, Moro MA, Dávalos A. A polymorphism in the *EAAT2* promoter is associated with higher glutamate concentrations and higher frequency of progressing stroke. J Exp Med. 2006;203:711–717.
12. Abbasi S, Rasouli M, Nouri M, Kalbasi S. Association of estrogen receptor-α A908G (K303R) mutation with breast cancer risk. Int J Clin Exp Med. 2013;6:39–49.
13. Hu F, Li D, Wang Y, Yao X, Zhang W, Liang J, Lin C, Ren J, Zhu L, Wu Z, Li S, Li Y, Zhao X, Cui B, Dong X, Tian S, Zhao Y. Novel DNA variants and mutation frequencies of hMLH1 and hMSH2 genes in colorectal cancer in the Northeast China population. PLoS ONE. 2013;8:e60233.
14. Janke T, Schwaiger K, Ege M, Fahn C, von Mutius E, Bauer J, Mayer M. Analysis of the fungal flora in environmental dust samples by PCR-SSCP. Meth Curr Microbiol. 2013;67:156–169.
15. Legrand S, Marque G, Blassiau C, Bluteau A, Canoy AS, Fontaine V, Jaminon O, Bahrman N, Mautord J, Morin J, Petit A, Baranger A, Rivière N, Wilmer J, Delbreil B, Lejeune-Hénaut I. Combining gene expression and genetic analyses to identify candidate genes involved in cold responses in pea. J Plant Physiol. 2013;170:1148–1157.
16. Choy YS, Dabora SL, Hall F, Ramesh V, Niida Y, Franz D, Kasprzyk-Obara J, Reeve MP, Kwiatkowski DJ. Superiority of denaturing high performance liquid chromatography over single-stranded conformation and conformation-sensitive gel electrophoresis for mutation detection in TSC2. Ann Hum Genet. 1999;63:383–391.
17. Gross E, Arnold N, Goette J, Schwarz-Boeger U, Kiechle M. A comparison of BRCA1 mutation analysis by direct sequencing, SSCP and DHPLC. Hum Genet. 1999;105:72–78.
18. Dobson-Stone C, Cox RD, Lonie L, Southam L, Fraser M, Wise C, Bernier F, Hodgson S, Porter DE, Simpson AH, Monaco AP. Comparison of fluorescent single-strand conformation polymorphism analysis and denaturing high-performance liquid chromatography for detection of EXT1 and EXT2 mutations in

hereditary multiple exostoses. Eur J Hum Genet. 2000;8:24–32.

19. Kakavas VK, Plageras P, Vlachos TA, Papaioannou A, Noulas VA. PCR-SSCP: a method for the molecular analysis of genetic diseases. Mol Biotechnol. 2008;38:155–163.
20. Ronaghi M, Uhlén M, Nyrén P. A sequencing method based on real-time pyrophosphate. Science. 1998;281:363–365.
21. Ahmadian A, Lundeberg J, Nyrén P, Uhlén M, Ronaghi M. Analysis of the p53 tumor suppressor gene by pyrosequencing. Biotechniques. 2000;28:140–144. 146-147.
22. Alderborn A, Kristofferson A, Hammerling U. Determination of single-nucleotide polymorphisms by real-time pyrophosphate DNA sequencing. Genome Res. 2000;10:1249–1258.
23. Nordström T, Ronaghi M, Forsberg L, de Faire U, Morgenstern R, Nyrén P. Direct analysis of single-nucleotide polymorphism on double-stranded DNA by pyrosequencing. Biotechnol Appl Biochem. 2000;31:107–112.
24. Ronaghi M, Elahi E. Pyrosequencing for microbial typing. J Chromatogr B Analyt Technol Biomed Life Sci. 2002;782:67–72.
25. Doostzadeh J, Shokralla S, Absalan F, Jalili R, Mohandessi S, Langston JW, Davis RW, Ronaghi M, Gharizadeh B. High throughput automated allele frequency estimation by pyrosequencing. PLoS ONE. 2008;3:e2693.
26. Akca H, Demiray A, Yaren A, Bir F, Koseler A, Iwakawa R, Bagci G, Yokota J. Utility of serum DNA and pyrosequencing for the detection of EGFR mutations in non-small cell lung cancer. Cancer Genet. 2013;206:73–80.
27. Pampliega O, Domercq M, Villoslada P, Sepulcre J, Rodríguez-Antigüedad A, Matute C. Association of an *EAAT2* polymorphism with higher glutamate concentration in relapsing multiple sclerosis. J Neuroimmunol. 2008;195:194–198.
28. Shin HE, Han SJ, Lee KS, Park JW. Polymorphism of the glutamate transporter protein *EAAT2* and migraine transformation into chronic daily headache. J Clin Neurol. 2011;7:143–147.
29. Rajatileka S, Luyt K, El-Bokle M, Williams M, Kemp H, Molnár E, Váradi A. Isolation of human genomic DNA for genetic analysis

from premature neonates: a comparison between newborn dried blood spots, whole blood and umbilical cord tissue. BMC Genet. 2013;14:105.

30. Sherry ST, Ward MH, Kholodov M, Baker J, Phan L, Smigielski EM, Sirotkin K. dbSNP: the NCBI database of genetic variation. Nucleic Acids Res. 2001;29:308–311.
31. Sheffield VC, Beck JS, Kwitek AE, Sandstrom DW, Stone EM. The sensitivity of single-strand conformation polymorphism analysis for the detection of single base substitutions. Genomics. 1993;16:325–332.
32. Gasser RB, Hu M, Chilton NB, Campbell BE, Jex AJ, Otranto D, Cafarchia C, Beveridge I, Zhu X. Single-strand conformation polymorphism (SSCP) for the analysis of genetic variation. Nat Protoc. 2006;1:3121–3128.
33. Cooper PC, Rezende SM. An overview of methods for detection of factor V Leiden and the prothrombin G20210A mutations. Int J Lab Hematol. 2007;29:153–162.
34. Shikanai T, Silverman ES, Morse BW, Lilly CM, Inoue H, Drazen JM. Sequence variants in the FcepsilonRI alpha chain gene. J Appl Physiol. 2002;93:37–41.
35. Hasegawa M, Nishiyama C, Nishiyama M, Akizawa Y, Mitsuishi K, Ito T, Kawada H, Furukawa S, Ra C, Okumura K, Ogawa H. A novel -66T/C polymorphism in Fc epsilon RI alpha-chain promoter affecting the transcription activity: possible relationship to allergic diseases. J Immunol. 2003;171:1927–1933.
36. Potaczek DP, Sanak M, Mastalerz L, Setkowicz M, Kaczor M, Nizankowska E, Szczeklik A. The alpha-chain of high-affinity receptor for IgE (FcepsilonRIalpha) gene polymorphisms and serum IgE levels. Allergy. 2006;61:1230–1233.
37. Shifman S, Kuypers J, Kokoris M, Yakir B, Darvasi A. Linkage disequilibrium patterns of the human genome across populations. Hum Mol Genet. 2003;12:771–776.
38. Dallaspezia S, Poletti S, Lorenzi C, Pirovano A, Colombo C, Benedetti F. Influence of an interaction between lithium salts and a functional polymorphism in SLC1A2 on the history of illness in bipolar disorder. Mol Diagn Ther. 2012;16:303–309.

39. Spangaro M, Bosia M, Zanoletti A, Bechi M, Cocchi F, Pirovano A, Lorenzi C, Bramanti P, Benedetti F, Smeraldi E, Cavallaro R. Cognitive dysfunction and glutamate reuptake: effect of *EAAT2* polymorphism in schizophrenia. Neurosci Lett. 2012;522:151–155.
40. Natkaniec M, Potaczek DP, Sanak M. Single-stranded conformation polymorphism (SSCP)-driven indirect sequencing in detection of short deletion. Mol Biol Rep. 2009;36:1545–1547.
41. Sivertsson A, Platz A, Hansson J, Lundegerg J. Pyrosequencing as an alternative to single-strand conformation polymorphism analysis for detection of N-ras mutations in human melanoma metastases. Clin Chem. 2002;48:2164–2170.
42. Unemo M, Olcén P, Jonasson J, Fredlund H. Molecular typing of Neisseria gonorrhoeae isolates by pyrosequencing of highly polymorphic segments of the porB gene. J Clin Microbiol. 2004;42:2926–2934.
43. Schentrup AM, Allayee H, Lima JJ, Johnson JA, Langaee TY. Genotyping the GGGCGG tandem repeat promoter polymorphism in the 5-lipoxygenase enzyme gene (ALOX5) by pyrosequencing assay. Genet Test Mol Biomarkers. 2009;13:361–365.
44. Fakruddin, Chowdhury A. Pyrosequencing an alternative to traditional Sanger sequencing. Am J Biochem Biotechnol. 2012;8:14–20.
45. Dinu I, Mahasirimongkol S, Liu Q, Yanai H, Sharaf Eldin N, Kreiter E, Wu X, Jabbari S, Tokunaga K, Yasui Y. SNP-SNP interactions discovered by logic regression explain Crohn's disease genetics. PLoS ONE. 2012;7:e43035.

CHAPTER 11

TITRATION-FREE MASSIVELY PARALLEL PYROSEQUENCING USING TRACE AMOUNTS OF STARTING MATERIAL

Zongli Zheng[1], Abdolreza Advani[2] , O¨ jar Melefors[2,3], Steve Glavas[2], Henrik Nordstro¨ m[2,3], Weimin Ye[1] , Lars Engstrand[2,3] and Anders F. Andersson[2,4],

[1] Department of Medical Epidemiology and Biostatistics, Karolinska Institutet,

[2] Swedish Institute for Infectious Disease Control, Solna,

[3] Department of Microbiology, Tumor and Cell Biology, Karolinska Institutet, Stockholm and

[4] Limnology/Department of Ecology and Evolution, Evolutionary Biology Centre, Uppsala University, Uppsala, Sweden

ABSTRACT

Continuous efforts have been made to improve next-generation sequencing methods for increased robustness and for applications on low amounts of starting material. We applied double-stranded library protocols for the Roche 454 platform to avoid the yield-reducing steps associated with single-stranded library preparation, and applied a highly sensitive Taqman MGB-probe-based quantitative polymerase chain reaction (qPCR) method. The MGB-probe qPCR, which can detect as low as 100 copies, was used to quantify the amount of effective library, i.e. molecules that form functional clones in emulsion PCR. We also demonstrate that the distribution of library molecules on capture beads follows a Poisson distribution. Combining the qPCR and Poisson statistics, the labourintensive and costly titration can be eliminated and trace amounts of starting material such as precious clinical samples, transcriptomes of small tissue samples and metagenomics on low biomass environments is applicable.

INTRODUCTION

Recent developments in DNA sequencing techniques allow millions of DNA molecules to be sequenced in parallel in a short time (1–5). However, the requirement of large quantities of starting material (e.g. 500 ng DNA in the latest Roche 454 Rapid library protocol) limits the possibilities to sequence trace amounts of DNA, such as ancient DNA, precious clinical samples, EST sequencing of small tissue samples or metagenomic samples from low biomass environments. Even in applications not limited by input DNA amounts, library quantification is a substantial source for uncertainty in sequencing outcome: a too high DNA-to-bead ratio leads to a significant fraction of non-readable beads with multiple DNA templates (mixed beads), while a too low ratio will result in an inadequate amount of beads and cannot take advantage of the full sequencing capacity.

Standard methods for quantification of libraries such as UV spectrophotometry and fluorometry require DNA amounts hundreds to thousands times the amount needed for the actual sequencing and cannot distinguish amplifiable from non-amplifiable molecules, the latter stemming from, for example, inefficient ligation of adapters or DNA damage. Alternative approaches were recently suggested; including SYBR Green quantitative polymerase chain reaction (qPCR) (6), 5′ universal template Taqman qPCR or digital PCR (7). However, SYBR Green qPCR requires an accurate and difficult estimation of the DNA library size distribution, and 5′ universal template Taqman qPCR gives a relatively large variation in quantification values (7), while digital PCR is not accessible in many laboratories (8). The Roche 454 protocol recommends an optimal amount of library to be determined empirically using the sequencing-titration assay for FLX Standard libraries and, later, the emulsion-titration assay for Titanium libraries. Both methods require labour-intensive and costly emulsion-PCR and enrichment procedures, as well as additional sample material.

In this study, we have developed a qPCR method based on a highly sensitive and precise Taqman MGB-probe. The MGB probe was designed to be complementary to the adapters used to construct two different types of libraries for the Roche 454 Titanium sequencing platform. One library consisted of the traditional A and B adapters on either end of the DNA template (9), and the other consisted of the same 'Y' adapter on both ends (similar to the Illumina GA library (8); http://www.seqanswers.com). This qPCR setup quantifies only effective library, i.e. DNA that will be amplified to form a functional clone in the emulsion PCR. We also demonstrate that Poisson statistics, with its parameter λ equalling the qPCR-measured input library DNA-to-bead ratio, can be used to predict enrichment percentage, a key index for sequencing performance.

MATERIALS AND METHODS

Using Poisson distribution to predict enrichment percentage

Nine previously sequenced FLX Standard libraries in our sequencing core facility were re-quantified by a Taqman MGB-probe-based qPCR. These libraries had previously been sequenced using the sequencing-titration assay, as part of the FLX Standard protocol, in four concentrations (0.5, 2, 4 and 16 DNA-to-bead ratios) as measured by RNA6000 BioAnalyzer chip (Agilent). Because the sequencing was performed without enrichment, and generally almost all DNA-carrying beads passed the key filter, enrichment percentage (percentage of DNA-carrying beads) can be calculated using the sequencing result metrics (enrichment % = key passed wells/raw wells ×100%). Apart from these nine libraries, there were 23 additional libraries subjected to the sequencing titration assay in our core facility, but these samples were no longer available at the time of the current study. However, their sequencing titration assay results were useful for exploring how the sequencing outcome metrics (proportions of Single, Mix or Dots reads) change over different enrichment percentages. To plot the enrichment percentages on the *Y*-axis and input DNA-to-bead ratio on the *X*-axis, which however could not be re-quantified by qPCR, we derived the input DNA-to-bead ratio from Poisson-transformed enrichment percentage [–log (1 – enrichment fraction)]. This was appropriate since the purpose here was not to explore the effect of input DNA-to-bead ratio on enrichment percentage. We also carried out a prospective evaluation of Poisson prediction by an emulsion-titration using newly prepared AB and Y libraries.

By Poisson distribution, the probability that there are exactly *k* DNA molecules on one library capture bead ($k = 0, 1, 2, \ldots$) is equal to:

$$f(k;\lambda) = \frac{\lambda^k e^{-\lambda}}{k!},$$

where λ is the input DNA-to-bead ratio. Therefore, the probability of having zero and one DNA molecule on one bead is $e^{-\lambda}$ and $\lambda \cdot e^{-\lambda}$, respectively, and the enrichment fraction is $1 - e^{-\lambda}$.

Deviation from Poisson distribution was tested using paired *t*-test comparing the difference, with 95% confidence intervals (CIs), between enrichment percentage observed from the titration sequencing and the enrichment percentage predicted by measured library concentration and Poisson distribution, after linearity transformation, ln(1 – enrichment percentage). Pearson's correlation coefficient between the observed and predicted enrichment percentage was calculated. When enrichment percentage is close to 100%, subtle enrichment differences will lead to large differences in DNA-to-bead ratio, we therefore excluded data when enrichment was > 95% (which is also far from the optimal DNA-to-bead ratio, see 'Discussion' section). This resulted in exclusion of 3 out of 36 data points. All data were analysed and plotted using the R software (10) (http://www.R-project.org.).

Construction of qPCR standards

We constructed a specific PCR product to be used as qPCR standards (available from the authors on request). One 1:1000 diluted stock FLX

Standard library (1 μl) from our sequencing core facility was used as template for amplification with 5 pmol of each emPCR primer, 1 × PCR buffer, 1.5 mM $MgCl_2$, 1 U Taq DNA polymerase (Invitrogen) and 200 μM dNTP each. The PCR product was cloned with the TOPO TA kit (Invitrogen). Several colonies were selected, transferred to 50 μl of water, and boiled at 95°C for 10 min. The lysate (1 μl) was used as template for colony PCR amplification using primers targeting the flanking region of the pCR4-TOPO vector ligation site. PCR products were visualized on 1% agarose gel and the cell lysate that generated an amplicon of about 360 bp (corresponding to an about 200 bp insert) was selected for Sanger sequencing using M13 forward primer. The result showed that the sequence was 202 bp long and contained one copy of adapter B and, therefore, one copy of qPCR probe complementary sequence. This cell lysate (1 μl) was then amplified using emPCR primers and purified with MinElute kit (Qiagen). The purified product was quantified using Qubit fluorescence quantification system (Invitrogen) and the number of molecules was calculated (=ng × 9.17 × 10^{11}/202). This DNA was 10-fold serially diluted and the concentrations from 10^7 to 10^2 copies were used as qPCR standards. The qPCR reaction mixture (20 μl) contained 1 × Taqman Fast Universal PCR Master Mix (Applied Biosystems), 900 nM of each emPCR primer (Invitrogen), 200 nM Taqman MGB-probe (Applied Biosystems) and 2 μl of standards or test samples. The qPCR was performed on the ABI 7900HT Fast System with the following conditions: fast cycling mode; 95°C for 20 s; 40 cycles including 95°C for 1 s and 60°C for 60 s. Fluorescence was detected during the extension step. Both standard and test samples were run in triplicates. It should be noted that, if quantifying a single-stranded library with this method, the readout must be multiplied by two to account for the double-stranded qPCR standards.

Library construction for FLX Titanium sequencing

We implemented a simplified library preparation protocol (8) that generates double-stranded DNA library using FLX Standard A and B adapters, but without biotin labelling on the B adapter. We designed a Taqman MGB-probe targeting the B adapter. We also implemented the concept of Illumina GA library and designed a Y adapter such that it is compatible with the Roche 454 Titanium platform and contains a complementary sequence at its 'B' branch. Genomic DNA was extracted from *Helicobacter pylori* strain HPAG1 (11) using DNeasy kit (Qiagen) according to the manufacturer's instructions. DNA was nebulized and selected for sizes between 300 and 800 bp by Solid Phase Reversible Immobilization (Agencourt) according to the Roche 454 protocol. One nanogram of DNA measured by Qubit Fluorometers (Invitrogen) was used for the downstream procedure. Polyallomer centrifuge tubes (Beckman Coulter) were used in all the experiments to minimize tube wall adsorption and denaturation of DNA (http://frstrauss.free.fr).

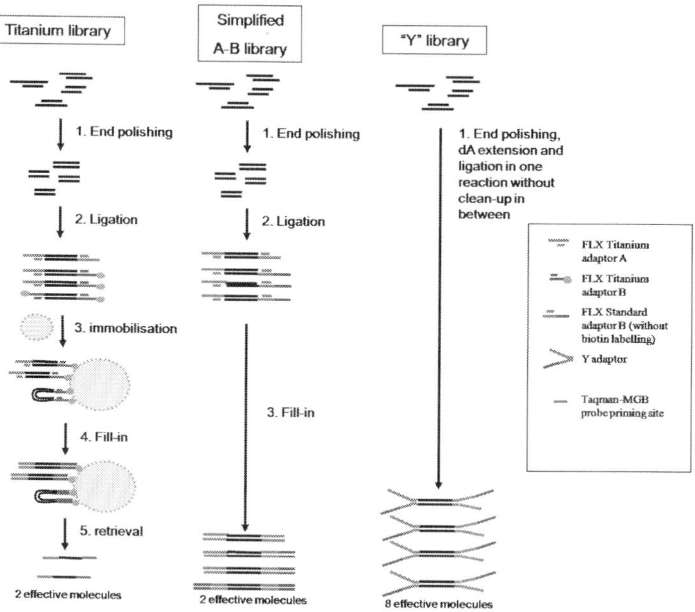

Figure 1: Schematic description of three types of library construction.

AB library

End polishing was conducted in 50 μl reaction volume with 1 ng of DNA, 1× End repair buffer, 5 μl end repair mix containing T4 DNA Polymerase and T4 PNK, 1 μl of 1 μM dNTPs (Enzymatics, MA, USA), and was incubated at 22°C for 30 min. After purification by AMPure beads (88 μl, 175% volume), DNA was eluted in 20 μl TE. Adapter ligation was conducted in a 25 μl reaction with 1× slow ligation buffer, 0.4 pmol of each of the A4 and B adapter (Invitrogen, USA) and T4

DNA Ligase 180 U (Enzymatics, MA, USA). We used the FLX Titanium adapter A4 (GS multiplex identifier number 4) because this adapter had performed well in previous experiments in our lab. The ligation reaction was incubated at 22°C over night, purified by AMPure beads (17.5 μl, 70% volume) and eluted in 25 μl TE. The eluted DNA was treated with 8 U Bst DNA polymerase Large Fragment (New England Biolabs), to fill in the 3′-junction nick between the adapter and sample DNA in a 30 μl reaction with 1× fill-in buffer, 30 μM dNTP and incubated at 37°C for 20 min. The double-stranded DNA library was purified with AMPure beads (21 μl, 70% volume) and eluted in 30 μl TE.

Y library

End polishing and 3′ dA extension was conducted in 22 μl reaction volume with 1 ng of DNA, 1× slow ligation buffer, 2.5 μl end repair mix containing T4 DNA polymerase and T4 PNK, 1 μl of 1 mM dNTPs (Enzymatics), 0.5 μl Klenow Fragment exo^{-} (New England Biolabs), 1× Taq polymerase buffer (Mg^{2+} free) and 0.5 μl Taq polymerase (Invitrogen). It was incubated at 12°C for 10 min, 37°C for 10 min, 72°C for 20 min and held at 4°C. Adapter ligation was conducted directly by adding 1 μl of 1 μM Y adapter, T4 DNA Ligase 180 U (Enzymatic) and was incubated at 12°C over night. The product was purified by AMPure XP beads (18 μl, 72% volume), and eluted in 30 μl TE. Immediately, not long, before library capture, the library was denatured into single-stranded, at 94°C for 2 min and held at 4°C, to avoid sequencing from both directions of a double-stranded template, which will result in mix reads.

It should be noted that this MGB probe is compatible with Roche FLX Standard library, not the FLX Titanium library nor the RL library, and that our AB and Y libraries are compatible with the FLX Titanium emPCR and sequencing kits.

RESULTS

Poisson behaviour of emulsion PCR

Optimizing the DNA-to-bead ratio in the emulsion PCR is a trade-off between minimizing the proportion of beads with multiple DNA templates and getting a sufficient amount of beads with template. The probability of getting zero, single or multiple DNA molecules per bead using different input DNA-to-bead ratios (λ) can be modelled using Poisson distribution. Since only DNA-carrying beads will be captured in the enrichment procedure, enrichment percentage corresponds to 1—probability of zero molecule per bead. Figure 2 shows theoretical enrichment percentage and probabilities of single and multiple templates per bead as functions of λ. When the DNA-to-bead ratio is small and covers the optimal range (0.08; see 'Discussion' section) it approaches a linear relationship with enrichment percentage (Figure 2B). In this range a 2-fold over- or underestimation of library concentration will all give satisfactory fractions of single-copy beads among enriched (98, 96 and 92% for ratios of 0.04, 0.08 and 0.16, respectively; Figure 2B). In contrast, if one would aim for a DNA-to-bead ratio of 1 (which should be avoided in practice) the same level of inaccuracy would result in 77, 58 and 31% single-copy beads for DNA-to-bead ratios of 0.50, 1.0 and 2.0, respectively.

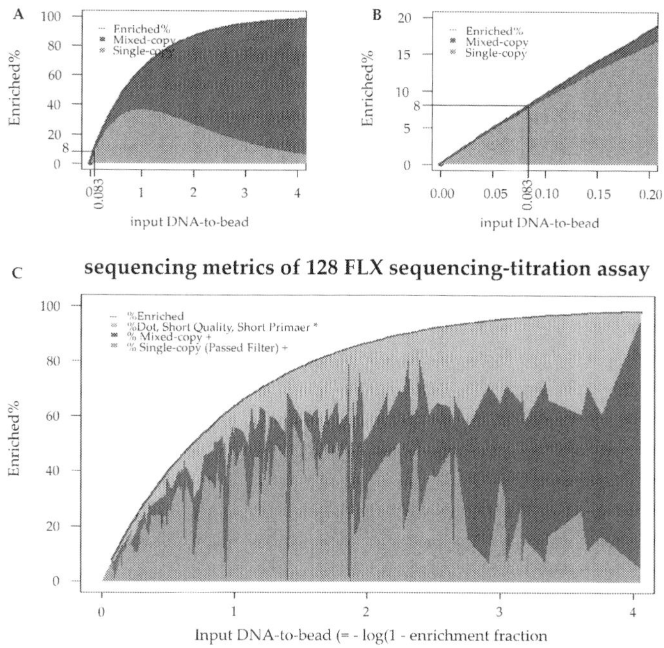

Figure 2: (**A**) Enrichment percentage, single copy and mixed copy DNAs in an emulsion drop/bead according to Poisson theory. In a subset (B), enriched

percentage as a function of input DNA-to-bead ratio approaches linearity when DNA-to-bead ratio decreases. The 8% enrichment percentage recommended by the Roche 454 protocol corresponds to a 0.083 DNA-to-bead ratio. (C) Sequencing outcome metrics as functions of the percentage of DNA-bearing beads (called enrichment percentage if enrichment were performed). The 40–60% enrichment percentage zone harbored a critical turning point, after which undesired beads becoming dominant quickly. As indicated by asterisks the Roche 454 Dot filter filters away those reads having too many negative flows due to poor incorporations or interruptions, the ShortQuality filter filters those reads failed the length test because of quality trimming and the ShortPrimer filter filters those failed the length test because of primer sequence trimming. As indicated by dagger the Roche 454 pipeline uses a positive flow percentage of 70% as a cut-off to distinguish Single- from Mixed-copy of templates.

We plotted the sequencing outcome metrics (single template and mixed template) as functions of enrichment percentage using the data from sequencing-titration assays of 32 FLX Standard libraries (four different DNA-to-bead ratios each) previously sequenced at our core facility. In accordance with the Poisson model, single template beads peaked at the range when the enrichment percentage was about 40–60; thereafter undesired beads quickly started dominating (Figure 2C).

Taqman MGB-probe-based qPCR for library quantification

In order to facilitate quantification of minute amounts of sequencing library, and to quantify only those library molecules that are effective, we designed a Taqman-MGB probe targeting the B adapter and the 'B' branch of the Y adapter. To estimate the precision and reproducibility of this Taqman-MGB probe setup, we re-quantified nine of the previously sequenced FLX Standard libraries (the remaining 23 libraries were not available). The libraries were 1:100 diluted and quantified with eight replicates each. The estimated concentrations of the diluted libraries were within the range of 10^5–10^8 molecules per microlitre. The mean (standard error) of the coefficient variation was 9.5% (1.6%). In a repeated run, these libraries were further diluted 1:10, spanning the range of working concentrations of 10^4–10^7, and quantified with triplicates; the mean (standard error) of the coefficient variation was 5.3% (1.5%).

Evaluating qPCR quantification and Poisson prediction retrospectively

We compared the qPCR quantification with previous quantifications done with BioAnalyzer. Large variations in enrichment percentages were observed for the same input DNA-to-bead ratios as measured by BioAnalyzer (Figure 3A), and the enrichment percentages were not predictable using Poisson (*P* value for the paired *t*-test comparing the observed and the predicted enrichment percentage was 0.0005, mean difference −42.2% (linearity transformation 3.77, 95% CIs 1.77–5.76). In contrast, when the libraries were measured with the qPCR method (Figure 3B), enrichment percentages were predictable using Poisson statistics: there was no significant difference between observed and predicted enrichment percentage (paired *t*-test *P* = 0.9157, mean difference 2.0% (linearity transformation −0.020, 95% CIs 0.365 to −0.500). Likewise, there was an improvement in the correlation between observed and predicted enrichment percentages using qPCR (Pearson's correlation coefficient = 0.68) as compared with BioAnalyzer (0.46).

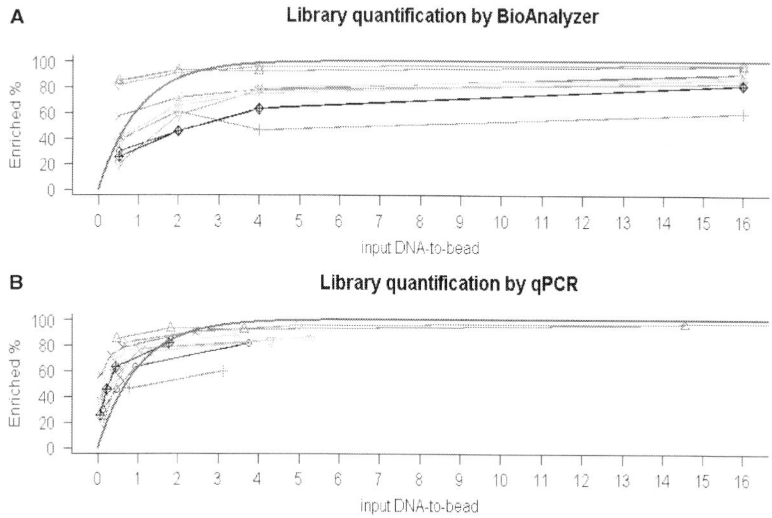

Figure 3: Nine stock libraries were previously quantified by BioAnalyzer as described in the Roche 454 Manual (A) and re-quantified by qPCR (B). Each colour indicates one library with four titration points. The blue curve indicates Poisson prediction of enrichment percentage. There was a deviation from Poisson prediction in (A) (P= 0.0005) and no deviation in (B) (P = 0.9157).

Evaluating qPCR quantification and Poisson prediction prospectively

To further evaluate how well MGB-probe qPCR and Poisson distribution can predict enrichment percentage, and also to demonstrate its applicability on the double-stranded libraries (AB library and Y library; Figure 1 and 'Materials and Methods' section), we performed emulsion-titration assays using newly prepared AB and Y libraries.

From 1 ng of fragmented DNA (mean size 500 bp), 1.15 million emPCR amplifiable AB library molecules and 53.6 million single-stranded Y library molecules were generated, respectively, as measured by qPCR (Figure 4A and C).

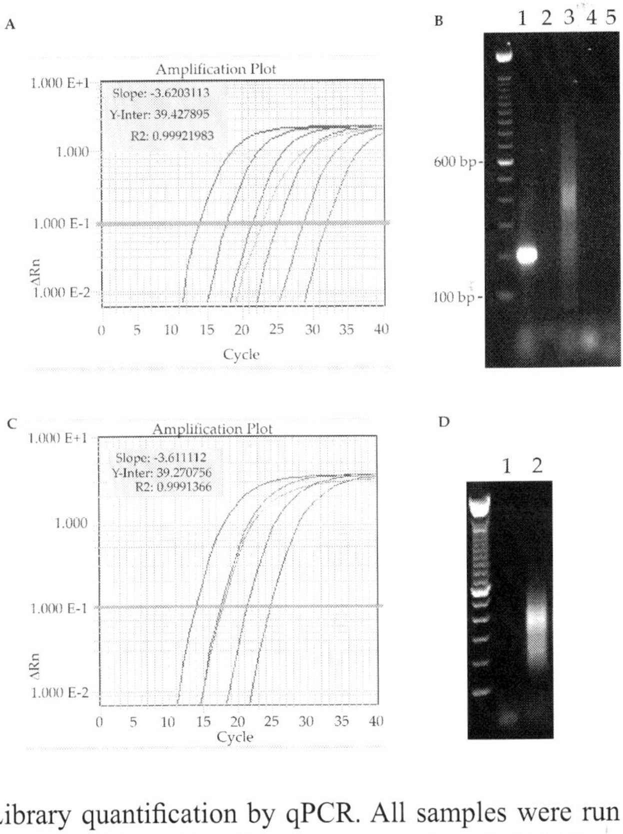

Figure 4: Library quantification by qPCR. All samples were run in triplicate and only the ones with median Ct values were plotted. (A) The six standards ranged from 107 to 102 copies. The tested AB library was 38 200 molecules/ml. The reactions containing the AB library template and one of the emPCR primers (A only or B only) generated no detectable fluoresces. (B) Gel electro-

phoresis of qPCR products. Lane 1: standard 104 copies; lane 2: no template control; lane 3: tested AB library; lane 4: the reaction containing the library template and primer A only; lane 5: the reaction containing the library template and primer B only. (C) The standards ranged from 107 to 104 copies. The tested Y library was 894 000 molecules/ml. (D) Gel electrophoresis of qPCR products. Lane 1: no template control; lane 2: Y library.

The qPCR amplification products were subsequently analysed by gel electrophoresis to make sure that the libraries had expected size distributions and to ensure that no adapter dimers had been carried over (Figure 4B and D). The putative dimers (84 bp for AB adapters and 79 bp for Y adapter) could not be observed among the amplified products. In addition, to test whether or not A–A and B–B tagged library molecules can be amplified, qPCR reactions containing the AB library templates and only one of the emPCR primers (either A or B) were performed. No detectable fluorescence (Figure 4A) and no PCR products visible on the gel were obtained (Figure 4B), indicating that the qPCR measurement corresponds to the amount of A-B tagged library, not B-B tagged library. The latter contains the probe complementary site and, if amplified, would interfere with the quantification.

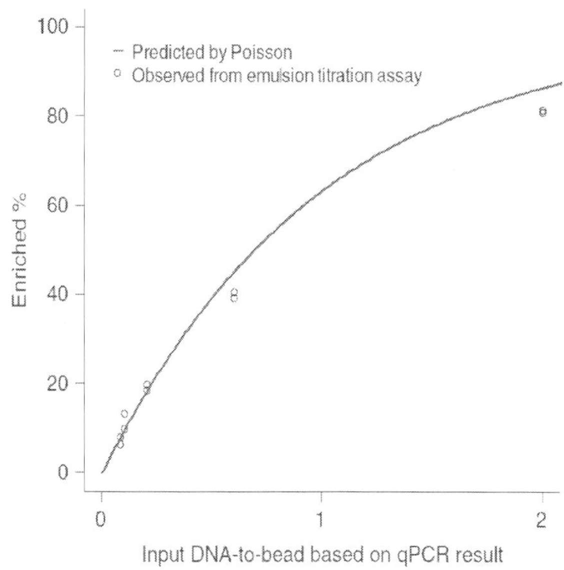

Figure 5: Enrichment percentage observed from emulsion-titration assay and predicted by Poisson distribution.

For the emulsion-titration assay, we used input DNA-to-bead ratios of 0.0796 for the AB library (see 'Sequencing outcome' below) and 0.1, 0.2, 0.6 and 2.0 for the Y library. According to Poisson prediction, the enrichment percentages should be 7.7, 9.51, 18.1, 45.1 and 86.5%. The actual enrichment percentages observed from the emulsion-titration assay for the five ratios (with duplicates) were 8.1 and 6.5%, 13.3 and 9.79%, 19.8 and 18.4%, 39.2 and 40.7%, and 81.3 and 81.0% (Figure 5). Hence, for these newly prepared libraries, observed enrichment fits Poisson prediction well (Pearson's correlation coefficient = 0.998 between predicted and observed enrichment).

Sequencing outcome

We set out to sequence the AB library using two M/S lanes (2 × 1/8 plate). The Roche 454 emPCR Kit supplies 4.8 million library capture beads for one M/S lane. We therefore used 10 μl of the AB library, corresponding to 0.382 million DNA molecules and hence a DNA-to-bead ratio of 0.0796. This was expected to generate 367195 (= 0.0765 × 4.8×10^6, where 0.0765 is the Poisson prediction of enrichment fraction $1 - e^{-0.0796}$) enriched beads. Assuming a 10% bead loss during the laboratory procedure, the remaining 90%, i.e. 330 475 enriched beads, would be nearly the amount needed according to the Roche 454 standard protocol (340 000). We processed two identical 10 μl of the library samples in parallel and thus generated two M/S lanes on a sequencing plate.

Following the enrichment process, the bead counter showed 389000 and 310000 beads for the two emPCR replicates. We loaded the enriched beads onto the sequencing plate and ran the sequencing according to the standard protocol. There were 104 091 and 96 151 filter-passed reads for the two lanes, falling well into the range of 80 000 to 120 000 reads per M/S lane of a successful sequencing run according to the Roche 454 protocol. Percentage of the filter-passed was 59.3% and 63.0%, median reads length 358 and 419 bp, and total bases 33.4 millions and 35.0 millions, respectively. Nearly all (99.3%) of the reads could be aligned to the *H. pylori* (strain HPAG1) genome and its plasmid using BLASTN with at least 30 bases matched and a significance value of 1E-6 (12), indicating that the contamination during library preparation, if any, was negligible. *De novo* assembly using Roche gsAssembler software resulted in a 1 584 532 bp genome, which was 99.3% the size of the reference genome HPAG1 (11).

DISCUSSION

With the Taqman-MGB probe-based qPCR and Poisson distribution, we were able to avoid the costly and labour-intensive titration assay. Using only one nanogram of fragmented DNA, we prepared enough library (1.15 million amplifiable AB library molecules and 53.6 millions Y library molecules, which is sufficient for 10 Titanium runs (~4 million enrichment beads to yield ~1 million high quality reads per run) for Roche 454 Titanium sequencing without the need for template pre-amplification by various means of whole genome amplification (13,14). However, it should be acknowledged that some factors causing sample loss remained, such as fragmentation of genomic DNA, low efficiency of ligation and a limited recovery in the enzymatic reaction clean-up (15). This was clearly shown by the much higher yield of Y library (sticky-end ligation) than AB library (blunt-end ligation and two additional reaction clean-up steps).

Library quantification by qPCR has earlier been proposed to overcome the lower detection limits of conventional methods (6,7). We used a Taqman MGB-probe to take advantage of its significantly higher sensitivity, specificity and reproducibility than conventional Taqman probes, at the same time only a shorter priming site is needed (16). This MGB probe rendered a precision (coefficient variation of 9.5%) comparable with digital PCR (11.8%), but better than a 5′ universal template Taqman probe (21.2%) (7).

We demonstrated that the distribution of DNA on beads after emulsification follows a Poisson distribution, which enabled us to predict enrichment percentage, a key index for successful sequencing. In general, the lower DNA-to-bead input, the lower percentage of beads with mixed templates and hence the more desirable results, provided that there are enough beads for sequencing loading. As a trade-off, a DNA-to-bead ratio close to 0.08 should be aimed at for optimal results (Figure 2). This is consistent with the Roche 454 protocol recommendation of 8% enrichment percentage (corresponds to a 0.083 DNA-to-bead ratio,

Figure 2B). The study using digital PCR by White *et al.* (7) used ratios in the range of 0.08 to 0.30, providing independent support for this DNA-to-bead ratio. Furthermore, the Poisson distribution demonstrates that imprecise library quantifications within 2-fold over- or under-estimations will all give satisfactory results when DNA-to-bead ratio is low, while this level of inaccuracy will lead to undesired results when DNA-to-bead ratio is high. In a previous study, a linear regression model was used to correlate input DNA-to-bead ratio with enrichment percentage (17). The linear regression method may suffer from limitations of a positive intercept (meaning that there will be enriched beads even with no input library) and unlimited enrichment percentage (can be higher than 100% when DNA amount increases). However, when the input DNA-to-bead ratio is low, prediction from a linear regression will approach that from a Poisson distribution and is then acceptable.

It should be noted that the physical capture of libraries on beads is performed before emulsification for FLX Standard libraries, and during emPCR for FLX Titanium libraries. The emulsification of the Roche 454 FLX Titanium platform is performed under tightly controlled conditions, such that aqueous 'microreactors' containing no more than a single bead are generated (18). Moreover, the total volume of the aqueous phase (DNA library) added for emulsification is strictly controlled so that almost all of the aqueous droplets contain one bead, i.e., few droplets contain no bead. Only when both of these two conditions are held, the distribution of DNA molecules on the beads can be expected to follow Poisson distribution. This explains why a limited volume of library (1–10 μl for small-scale and <100 μl for large-scale emPCR, as described in the Roche 454 protocol) should be used for good results.

Our results confirmed a previous study that the AB library (a mixture of A–B, A–A and B–B tagged templates, without the need to clean out the latter two) are suitable for Roche 454 sequencing (9). The A–A and B–B tagged library templates could not be amplified in PCR. This is because hairpin structures, formed in the annealing process by the complementary sequences on either end of the single-stranded DNA, can prevent annealing of primers because: (i) the much higher local annealing temperature of the hairpin structures (>80°C for both the 40 bp 'A hairpin' and 44 bp 'B hairpin') than that of the PCR primers (60°C) and (ii) the complementary sequences in the ends of the same single-stranded DNA molecule are closer and more accessible for binding than surrounding primers at normal concentrations. This was confirmed by the observation that there was no detectable fluorescence and no visible PCR products generated in the qPCR or in a normal no-probe PCR reactions containing primer A only or primer B only. Interestingly, a recent study using transposons to generate shotgun libraries showed that A–A or B–B tagged templates generated ~1:1000 the signal of A–B tagged templates (19). The fact that their A–A or B–B templates were actually amplified (although weakly) could be due to a weaker amplification inhibitory effect associated with their shorter hairpins than with ours.

The qPCR amplification slopes of the libraries were lower than those of the standards, indicating that portions of the libraries, despite being A–B tagged, were not well amplified. Similarly in the emPCR of the FLX platform, not every DNA in the emulsion droplet will be well amplified (18). Because short amplicons are generally amplified better than long ones, we used small sizes (202 bp) as qPCR standards to quantify effective library. In contrast, a standard of size similar to the library median (e.g. 500 bp) would lead to quantification of total library rather than effective library and, consequently, an overestimated DNA-to-bead ratio. Thus, if one aims for a 0.08 DNA-to-bead ratio, a quantification based on 500-bp standards would result in too few beads being enriched. In addition, to mimic the long extension time in the emPCR, the qPCR extension time was prolonged from the normal 20 seconds to 60 seconds. If the amplicons are not in the expected size range, as seen from the gel electrophoresis, a longer extension such as an addition of 68°C for 60 s in the qPCR cycling programme may be needed.

During the revision of this manuscript, Roche released a simplified library protocol, 'Rapid', in which a starting amount of DNA of at least 500 ng is recommended. This is an improvement compared to previous protocols, but for applications where limited amounts of material are available this may still be too much. The qPCR quantification and Poisson prediction methods presented here for the Roche 454 sequencer are expected to be useful also for the SOLiD, Solexa and Ion Torrent sequencing platforms. However, our methods have been validated on one bacterial species. For sequencing of more complex samples such as metazoan or plant genomes, further validation studies might be needed.

ACKNOWLEDGEMENTS

The authors acknowledge the suggestions from one of the reviewers that improved the manuscript. They also thank Wilhelm Paulander for help with cloning; Hedvig Engstrom Jakobosson for help with qPCR; Lena Eriksson for sample preparation and Sten Linnarsson for discussion on handling of trace amounts of DNA.

REFERENCES

1. Harris TD, Buzby PR, Babcock H, Beer E, Bowers J, Braslavsky I, Causey M, Colonell J, Dimeo J, Efcavitch JW, et al. Single-molecule DNA sequencing of a viral genome. Science. 2008;320:106–109.
2. Bentley DR, Balasubramanian S, Swerdlow HP, Smith GP, Milton J, Brown CG, Hall KP, Evers DJ, Barnes CL, Bignell HR, et al. Accurate whole human genome sequencing using reversible terminator chemistry. Nature. 2008;456:53–59.
3. Shendure J, Porreca GJ, Reppas NB, Lin X, McCutcheon JP, Rosenbaum AM, Wang MD, Zhang K, Mitra RD, Church GM. Accurate multiplex polony sequencing of an evolved bacterial genome. Science. 2005;309:1728–1732.
4. Margulies M, Egholm M, Altman WE, Attiya S, Bader JS, Bemben LA, Berka J, Braverman MS, Chen YJ, Chen Z, et al. Genome sequencing in microfabricated high-density picolitre reactors. Nature. 2005;437:376–380.
5. Clarke J, Wu HC, Jayasinghe L, Patel A, Reid S, Bayley H. Continuous base identification for single-molecule nanopore DNA sequencing. Nat. Nanotechnol. 2009;4:265–270.
6. Meyer M, Briggs AW, Maricic T, Hober B, Hoffner B, Krause J, Weihmann A, Paabo S, Hofreiter M. From micrograms to picograms: quantitative PCR reduces the material demands of high-throughput sequencing. Nucleic Acids Res. 2008;36:e5.
7. White RA, 3rd, Blainey PC, Fan HC, Quake SR. Digital PCR provides sensitive and absolute calibration for high throughput sequencing. BMC Genomics. 2009;10:116.
8. Linnarsson S. Recent advances in DNA sequencing methods—

general principles of sample preparation. Exp. Cell Res. 2010 doi:10.1016/j.physletb.2003.10.071.

9. Wiley G, Macmil S, Qu C, Wang P, Xing Y, White D, Li J, White JD, Domingo A, Roe BA. Methods for generating shotgun and mixed shotgun/paired-end libraries for the 454 DNA sequencer. Curr. Protoc. Hum. Genet. 2009 Chapter 18, Unit18.1.
10. R Development Core Team. R: A Language and Environment for Statistical Computing. Vienna, Austria: R Foundation for Statistical Computing; 2009.
11. Oh JD, Kling-Backhed H, Giannakis M, Xu J, Fulton RS, Fulton LA, Cordum HS, Wang C, Elliott G, Edwards J, et al. The complete genome sequence of a chronic atrophic gastritis *Helicobacter pylori* strain: evolution during disease progression. Proc. Natl Acad. Sci. USA. 2006;103:9999–10004
12. Altschul SF, Madden TL, Schaffer AA, Zhang J, Zhang Z, Miller W, Lipman DJ. Gapped BLAST and PSI-BLAST: a new generation of protein database search programs. Nucleic Acids Res. 1997;25:3389–3402.
13. Blow MJ, Zhang T, Woyke T, Speller CF, Krivoshapkin A, Yang DY, Derevianko A, Rubin EM. Identification of ancient remains through genomic sequencing. Genome Res. 2008;18:1347–1353.
14. Pinard R, de Winter A, Sarkis GJ, Gerstein MB, Tartaro KR, Plant RN, Egholm M, Rothberg JM, Leamon JH. Assessment of whole genome amplification-induced bias through high-throughput, massively parallel whole genome sequencing. BMC Genomics. 2006;7:216.
15. Maricic T, Paabo S. Optimization of 454 sequencing library preparation from small amounts of DNA permits sequence determination of both DNA strands. Biotechniques. 2009;46:51–52, 54–57.
16. Kutyavin IV, Afonina IA, Mills A, Gorn VV, Lukhtanov EA, Belousov ES, Singer MJ, Walburger DK, Lokhov SG, Gall AA, et al. 3′-minor groove binder-DNA probes increase sequence specificity at PCR extension temperatures. Nucleic Acids Res. 2000;28:655–661.
17. Sandberg J, Stahl PL, Ahmadian A, Bjursell MK, Lundeberg J.

Flow cytometry for enrichment and titration in massively parallel DNA sequencing. Nucleic Acids Res. 2009;37:e63.

18. Roche Diagnostics GmbH. GS FLX Titanium emPCR Method Manual. Roche Applied Science, 68298 Mannheim, Germany: Roche Diagnostics; 2008.
19. Syed F, Grunenwald H, Caruccio N. Optimized library preparation method for next-generation sequencing [advertising feature] Nat. Methods. 2009;6:i–ii.

CHAPTER 12

PYROSEQUENCING ON A GLASS SURFACE

Ana V. Almeida [a,b,d], **Andreas Manz** [a,b,d] **and Pavel Neužil**[a,c,d]

[a]KIST-Europe, Microfluidics Group, Campus E7.1, 66111 Saarbrücken, Germany.

[b]Mechatronics Department, Universität des Saarlandes, Campus A5, Saarbrücken, Germany

[c]Department of Microsystem Engineering, School of Mechanical Engineering, Northwestern Polytechnical University (NPU), 127 West Youyi Road, Xi'an Shaanxi, PR China

[d]Central European Institute of Technology (CEITEC), Brno University of Technology (BUT), Technická 3058/10, Czech Republic

ABSTRACT

We demonstrate the use of open-surface microfluidics to sequence DNA by pyrosequencing at the plain hydrophobically coated surface of a microscope glass cover slip. This method offers significant advantages in terms of instrument size, simplicity, disposability, and functional integration, particularly when combined with the broad and flexible capabilities of open-surface microfluidics. The DNA was incubated on superparamagnetic particles and placed on a hydrophobically coated glass substrate. The particles with bound DNA were moved using magnetic force through microliter-sized droplets covered with mineral oil to prevent water evaporation from the droplets. These droplets served as reaction "stations" performing pyrosequencing as well as washing stations. The resequencing protocol with 34-mer single-stranded DNA (ssDNA) was used to determine the reaction performance. The de novo sequencing was performed with 51-mer and 81-mer ssDNA. The method can be integrated with previously shown sample preparation and PCR into a single sample-to-answer system on a plain glass surface.

INTRODUCTION

Deoxyribonucleic acid (DNA) molecules carry the majority of genetic information for all living species as well as viruses. This information is coded in a sequence of nucleotides in the DNA[1] and the process to determine the precise order of nucleotides is referred to as sequencing.[2] Classical Sanger sequencing is based on selective incorporation of chain-terminating dideoxynucleotides by DNA polymerase during in vitro DNA replication.[3,4] This original method, as well as others developed later, has become an essential tool for many basic and applied research applications.[5] Sequencing instruments were developed during the mapping of the entire human genome, which was completed in 2003 (ref. 6). Since then, a number of methods have been implemented in commercial DNA sequencers[2,7] and they are now capable of rapidly sequencing the whole genome. These are routinely used in microbial,[8–10] DNA methylation,[11] and single nucleotide polymorphism (SNP) analyses.[12] However, these systems are complex, typically bulky, and require qualified personnel to operate them.

A frequently used sequencing method is pyrosequencing. This is a real-time sequencing-by-synthesis approach based on the detection of the inorganic pyrophosphate (PPi) released during the DNA polymerization reaction.[13–15] This cascade of three enzymatic reactions is shown below:

$$(\mathrm{DNA})_n + \mathrm{dNTP} \xrightarrow{\text{DNA poymerase}} (\mathrm{DNA})_{n+1} + \mathrm{PPi}, \tag{1}$$

$$PPi + APS \xrightarrow{\text{ATP sulphurylase}} ATP + SO_4^{-2} \tag{2}$$

$$ATP + \text{luciferin} + O_2 \xrightarrow{\text{luciferase}} AMP + PPi + \text{oxyluciferin} + CO_2 + h\nu, \tag{3}$$

where APS stands for adenosine 5′-phosphosulphate, ATP for adenosine triphosphate, h is Planck's constant, ν is frequency of emitted photons during the bioluminescent reaction, and hν is its energy. The single-stranded DNA (ssDNA) is hybridized to a sequencing primer that can be immobilized on a substrate, such as superparamagnetic particles (SPP) via a biotin–streptavidin bond. In the presence of DNA polymerase, nucleotides (dNTP) can be sequentially introduced to the system and incorporated when complementary to the template strand. The resulting amount of photons released is proportional to the amount of ATP molecules generated. Unreacted nucleotides as well as excess ATP can be removed by washing before the next nucleotide addition.[16,17] The sequence of the template can be determined as the added nucleotide is known. A disadvantage of this method is its short read length of between 25 and 100 base pairs (bps) per template molecule.[18]

Microfluidics emerged as a powerful new field in 1993, following its demonstration in miniaturized capillary electrophoresis[19]and then by a flow-through polymerase chain reaction (PCR).[20] Nevertheless, both simple microfluidics systems required vast external instrumentation. Gradually, the instrumentation was optimized and new, simplified approaches developed[21,22] and microfluidics-based systems started to play an increasingly important role in addressing the demand for faster, more accessible and easier methods for applications, including DNA sequencing technologies.[23–25] Traditionally, microfluidics devices were based on closed-channel systems fabricated by lithography, etching, and bonding processes. The process was complex and devices suffered from the formation of air bubbles and their trapping. Motivated by advancements in point-of-care (POC) and home-care systems, open-surface microfluidics (OSM) for biotechnology has been recently gaining attention.[26] This technique requires simpler fabrication than traditional microfluidic chips, which eliminates cleanroom processes, suppresses air bubble-related problems, and offers user-friendly operation due to the device simplicity.

Sample manipulation and preparation were demonstrated using magnetic particles in a closed system by integrating various functions from a biological sample to DNA extraction and purification.[27] The same magnetic particle manipulation was used with OSM, where undesirable water evaporation was eliminated by sample encapsulation with mineral oil, forming a "virtual reaction chamber" (VRC). A proof-of-concept of this type of microfluidics was demonstrated by PCR and sample preparation strategies.[28–30] A number of complex bioassays have been shown where droplets were self-contained and operated as reaction chambers as well as sample transportation units.[31] The mineral oil encapsulation effectively prevented sample evaporation for hours even at elevated temperature.[28] Extreme cases were demonstrated earlier with water samples heated to ≈200 °C (ref. 32 and 33), where no water evaporation from the encapsulated sample was observed.

Here, we present an approach to perform three-enzyme pyrosequencing using magnetically actuated VRCs by means of the OSM concept. The reactions are conducted on a plain microscope glass cover slip, thereby eliminating all microfabrication steps involved in other microfluidic approaches. These are performed on a disposable substrate, which makes the platform significantly simpler than conventional microfluidics with closed channels and pumps, OSM-type with surface acoustic waves (SAW),[34,35] or using electro-wetting techniques.[36]

MATERIALS AND METHODS

Materials

For pyrosequencing preparation, ≈100 µL of paramagnetic particles M280 Dynabeads from Life Technologies, GmbH (Germany) was washed three times and resuspended in 100 µL of binding buffer (≈10 mM Tris buffer pH ≈7.6, ≈2 M NaCl, ≈1 mM EDTA solution and ≈0.1% Tween 20). All chemicals were provided by Sigma-Aldrich, GmbH (Germany). Amounts of ≈4 µg of biotinylated PCR amplicon in ≈100 µL of H_2O were added to the particles and incubated at ≈65 °C for ≈15 min with periodic mixing for binding of the DNA. The double-stranded amplicon was denatured by exposing the beads to ≈100 µL of ≈0.5 M NaOH for ≈1 min. The beads were then washed by NaOH followed by three additional washing steps in magnesium annealing buffer (≈20 mM Tris acetate, pH ≈7.6, ≈5 mM magnesium acetate) to obtain ssDNA. The beads were resuspended in

≈100 μL of magnesium annealing buffer prior to adding ≈5 μL of ≈10 μM sequencing primer for hybridization of the primer. The solution was heated to ≈80 °C for ≈2 min and cooled back to room temperature using a heat block to allow the primer to hybridize. The bead/template/primer complex was washed three times in pyrosequencing wash buffer as described below; then, ≈15 μg of single-stranded binding protein (SSB) by Promega, GmbH (Germany) was added and the mixture incubated at room temperature for ≈10 min. The beads were washed with pyrosequencing wash buffer and resuspended in ≈200 μL of pyrosequencing wash buffer to achieve a final bead mass concentration of ≈5 μg μL^{-1}. On the basis of the manufacturer's specified binding capacity, it is estimated that the ≈5 μL aliquot used in each sequencing reaction contained between ≈150 and ≈460 pmol of template DNA.

Separate stock solutions were prepared for pyrosequencing: washing buffer, enzymes, dNTPs, and DNA. Pyrosequencing washing buffer was prepared by mixing 100 mM Tris acetate pH ≈7.6 with ≈0.5 mM ethylenediaminetetraacetic acid (EDTA) pH ≈ 8.0, ≈5 mM magnesium acetate, and ≈0.01% Tween 20. Enzyme solution was prepared with pyrosequencing washing buffer by adding ≈13.5 mU μL^{-1} ATP sulfurylase by New England BioLabs Inc. (Germany), ≈1.5 μg μL^{-1} luciferase and ≈1.8 μg μL^{-1}D-luciferin by Sigma-Aldrich, GmbH (Germany), ≈6.9 U μL^{-1} Klenow (exo-) fragment by ThermoFisher Scientific, GmbH (Germany), and ≈0.01% Tween 20, ≈18 μM stock solutions of each dNTP by Thermoscientific, GmbH (Germany) was added to pyrosequencing washing buffer with ≈15 μM APS by Sigma-Aldrich (Germany), ≈3 mM dithiothreitol (DTT) by Sigma-Aldrich, GmbH (Germany), ≈120 ng μL^{-1} single-stranded binding protein (SSB), and ≈4.6 U μL^{-1} Klenow (exo-) fragment.

Re-sequencing. We used a synthetic biotinylated 55-mer sequence B-5′-GAT GAC TGT AAGG GGA GTC AAG GTG CAC CTT TAG ACT GCC GTC GTT TTA CAA C-3′ and the sequencing primer 5′-GTA AAA CGA CGG CCA GT-3′.

***De novo* sequencing.** Using nested PCR for biotinylation of the amplicon (amplicon length 221 bps), a sequence from the pfcrt 271 region from human genomic DNA was selected. (Primers: external forward: 5′-GGC TAT GGT ATC CTT TTT CCA A-3′; external reverse: 5′-CGA CTG TGT TTC TTC CCA AG-3′; internal forward: B-5′-ATC

CTT TTT CCA ATT GTT CAC TTC-3′; and internal reverse: 5′-CGA AAC CAT TTT TTA TAT TTG TCC-3′.) Sequencing was performed using the sequencing primer 5′-TTT CCT AAT TAA TTC TTA CG-3′.

Long-read sequencing. We used a synthetic biotinylated 81-mer sequence B-5′-GTT TAC TCA TAT ATA CTT TAG ATT GAT TTA AAA CTT CAT TTT TAA TTT AAA AGG ATC TAG GTG AAG ATC CTT TTT GAT AAT CCT TTA AGA CTG CCG TCG TTT TAC AAC-3′ and the sequencing primer 5′-GTA AAA CGA CGG CAG TC-3′.

All primers and synthetic sequences were purchased from Eurofins MWG Operon, GmbH (Germany).

Methods

Prior to pyrosequencing, DNA has to be extracted from the sample, purified, and amplified by PCR. Most of these steps are either labor-intensive or require complex robotic systems. Except for pyrosequencing, all other necessary steps have been previously integrated using the OSM concept.[29,30] The system proposed here is fully compatible with previous work, thus, it is possible to be integrated later on into a single "sample-to-answer" device.

The pyrosequencing experiments were conducted on a glass cover slip as shown schematically in Fig. 1A. The ssDNA bound to SPP was placed into a first station as shown in (a), with a magnet underneath it. The particles are then dragged (b) using magnetic force into the first (E) enzyme-containing droplet. This droplet is in close vicinity with the (S + G) substrate and nucleotide dGTP-containing droplet, thus, the droplets merged and the pyrosequencing reaction sequence was initiated. The particles are then dragged into the washing droplet. This washing droplet is at a larger distance from the other droplets, so no merging of droplets occurs. The washing steps were designed as the major determinant of maximum read length via complete removal of unincorporated or excess nucleotides at each step. We have proposed washing the bead/template/primer complexes using washing buffer to remove nucleotides as well as other reaction byproducts between nucleotide additions. Subsequently, the particles are dragged into another station (c) with two droplets, one again with enzyme and the other one with substrate and another nucleotide (dTTP).

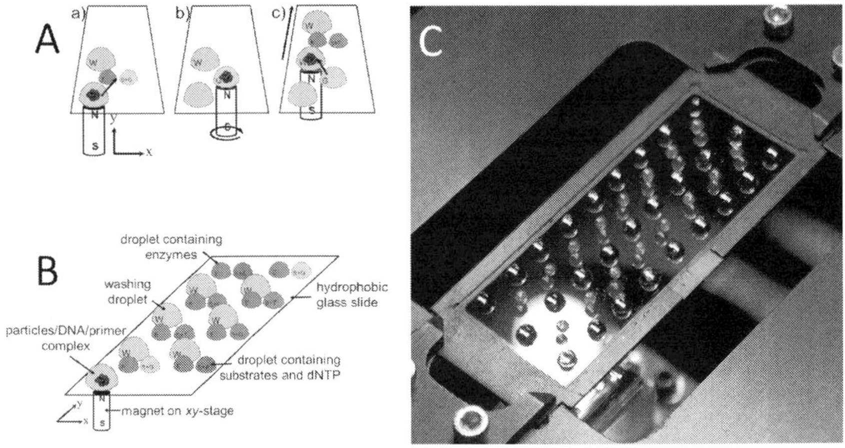

Figure. 1 (A) Representation of a typical nucleotide addition cycle. All droplets are covered with oil, thereby effectively preventing evaporation of water from the sample as well as reagent degradation by oxidation. (a–c) Three phases of the pyrosequencing reaction. (B) Principle of the experiment setup; (W) washing droplets; (E) droplets containing all the needed enzymes for the pyrosequencing reaction (S + G − T − C − A). (C) Photograph of the setup consisting of the hydrophobically coated glass cover slip mounted onto a motorized X–Y translation stage with a neodymium–iron–boron magnet fixed beneath the cover slip. The droplets contained color ink to increase the contrast.

A glass containing eight reaction stations is schematically shown in Fig. 1B and a photograph of an actual setup with 16 stations in Fig. 1C as well as a single-loading and unloading droplet containing DNA/primer/SPP. The droplets were colored with ink to increase the contrast. The glass with the dimension of $\approx 60 \times \approx 20$ mm and thickness of ≈ 170 μm was treated with 1H,1H,2H,2H-perfluorodecyltriethoxysilane (FAS17), making the surface highly hydrophobic, with a water contact angle of $\approx 95°$. The glass cover slip was placed on a computer controlled X–Y stage. A disc-shaped neodymium magnet with a diameter of ≈ 3 mm, model N45 mm by Supermagnetic, GmbH (Germany) was placed underneath the glass cover slip.

The DNA/primer/SPP complex was immobilized on SiO_2-covered SPP and moved through the individual reaction/washing stations in the form of droplets using magnetic force. These "stations" were formed by pipetting individual droplets onto the glass covered with a mineral oil M5904 by Sigma-Aldrich, GmbH (Germany), thereby preventing water evaporation from reaction droplets. After a series of experiments, the volumes of stations as well as oil covering them are summarized in Table 1.

Table 1 Stations and covering oil volumes

	Washing droplet	**Enzyme droplet**	**Substrate and nucleotide droplet**
Droplet volume (μL)	≈10	≈3	≈2
Oil volume (μL)	≈2	≈1	≈1

We placed a photomultiplier tube (PMT) model H10492-001 from Hamamatsu Photonics K.K. (Japan) above the glass aligned with the magnet underneath. Since the sample was always attracted by the magnet and kept on the glass above it, it was also directly below the PMT; thus, all reactions were conducted at the sample location with respect to the PMT. With this setup, both the magnet and the PMT were stationary while the glass was movable. The stage with the PMT and magnet was placed in a black box to eliminate ambient light noise. The PMT was equipped with an internal current-to-voltage converter, with its gain adjusted to ≈0.7 V using a custom made PMT controller setting the gain value to $\approx 10^5$ as interpolated from a graph supplied by the manufacturer. The PMT output was directly connected to the input of an oscilloscope model DPO 7054 provided by Tektronix, Inc. (USA) to record optical power as a function of time.

The PMT voltage output as function of time captured by an oscilloscope was processed using a custom written MATLAB script. First, the data histogram based on frequency data count was formed and its peak was used to determine the signal baseline, which was subsequently subtracted from the data. All peaks were identified and individually processed to extract the values of maximum amplitude as well as their areas under the curve (AUC). The AUC unit corresponds to the electrical charge C, as it is time integration of electrical current i from 0 s to t s:

$$\text{AUC} \approx \text{const}\int_0^t i\,dt, \qquad (4)$$

where const is a conversion factor. It consists of a gain of PMT and its transconductance amplifier and PMT efficiency of ≈10^5, ≈1 μA V^{-1}, and ≈85 mA W^{-1}, respectively for light wavelength of ≈550 nm.

RESULTS AND DISCUSSION

An initial series of experiments with and without a washing step was conducted. Performing the pyrosequencing reactions without the washing step resulted in a 30% carry-over signal by the DNA/primer/SPP complex. Once the washing step was incorporated, the carry-over signal was eliminated; thus, each reaction was terminated with a single washing step. As one can see from the results, once the washing step was incorporated into the experiment design, no emitted light was observed from the sample, which allows us to conclude that no significant amount of chemicals were carried over.

For each experiment, we used new (hydrophobically coated) glass cover slips to eliminate potential sample-to-sample cross contamination, which is an advantage of having a disposable and microfabrication-free platform.

We also tried to overcome the diffusion-dependent rate of the pyrosequencing process by moving SPP within the droplet during the reaction. The stage moved by ±≈2 mm along both X and Y axes at the speed of ≈2 mm s^{-1}, which shortened the reaction time by a factor of almost 2. Initially, the signal peak amplitude was ≈1.00 V and its area was ≈0.32 mC. Stimulated by stirring, the peak amplitude increased to ≈1.99 V and the area dropped to ≈0.27 mC. The AUC amplitude dropped to ≈85% of its original value while the signal peak amplitude increased by factor of ≈2. The signal reached the baseline amplitude at time ≈275 s; thus, we decided that keeping the reaction under the PMT for ≈300 s should be sufficient.

Next, we performed a set of experiments to determine optimum DNA. We used ≈55 μg of beads and DNA concentration from ≈0.3125 pmol to ≈20 pmol. The DNA concentration of ≈1.25 pmol seemed to be optimum. This concentration resulted in well-distinguished groups for each number of nucleotide incorporation. The peak voltage and AUC amplitude were (0.969 ± 0.227) V and (0.147 ± 0.021) mC (mean ± standard deviation) per single nucleotide incorporation, respectively, with correlation coefficient of 0.959 and 0.896 for peak voltage and AUC, respectively. This peak voltage signal was ≈100 times higher than the background noise while requiring minimum DNA consumption.

As mentioned earlier, each glass cover slip can accommodate up to 16 reaction sets (washing + enzyme and substrate droplets). Sequencing of more than 16 nucleotides then requires multiple glass cover slip and the sample has to be unloaded from one glass and loaded to another glass. These loading/unloading stations were designed not to contain matching nucleotides, thus, they also served as reference points to determine the background signal. We pipetted out the particles from the unloading station and transferred them to another glass slip, with the aid of a magnet at the surface of the pipette tip, into a loading station. Once the experiments were completed, we calculated the AUC and peak voltage amplitudes from 5 transfers in unloading and loading stations and compared them with each other. We found that the loading PMT peak and AUC amplitude were (0.79 ± 0.08) V and (0.094 ± 0.005) mC (mean ± standard deviation), respectively. Unloading PMT and AUC amplitude were (0.79 ± 0.11) V and (0.098 ± 0.014) mC (mean ± standard deviation), respectively. The results during loading and unloading are practically identical, proving successful beads transfer. This information is crucial to verify that we can transfer beads from glass to glass while maintaining the same amplitude of emitted light due to nucleotide incorporation during DNA sequencing.

We also tested the maximum number of homopolymeric nucleotide additions per single incorporation. We observed that the amplitude of emitted light starts to drop after incorporations of more than seven nucleotides. We extracted the peak voltage and AUC amplitude of (1.451 ± 0.067) V and (0.175 ± 0.020) mC (mean ± standard deviation), respectively, per single nucleotide incorporation. Peak voltage amplitude provided more consistent signal than the one of AUC. The signal level after each successive wash cycle could also be diminished in the reaction complex because of the loss or denaturation of either the DNA template or the sequencing primer.

The system was calibrated with DNA concentration of ≈1.25 pmol by performing a re-sequencing experiment (Fig. 2). A glass cover slip was prepared with the dNTPs placed in predetermined order, which were incorporated into the complementary DNA strand resulting in a final 34-mer. The sequence consisted of 21 single, 3 double, 1 triple, and 1 quadruple nucleotide incorporations, and 4 mismatches. We had to use two glass cover slips with single sample transfer from the first glass to the second glass.

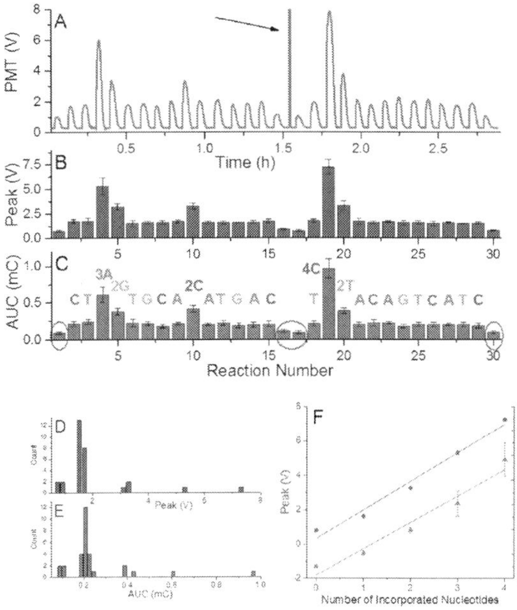

Figure 2: (A) A typical pyrogram or resequencing. (B) Mean value of peak voltage and its standard error as function of reaction number. A mismatched event was performed in the end of the run to determine the background emission (marked by blue ellipses). (C) Mean value of AUC and its standard error as the function of the reaction. Frequency counts of average peak values (D) and AUCs (E) distinguish five groups according to the number of incorporated nucleotides with a nearly perfect linear relationship. (F) Peak amplitude (blue circles – left axis) and AUC (red triangles – right axis) as a function of the number of incorporated nucleotides with slope of (1.657 ± 0.140) V and (0.214 ± 0.026) mC for peak voltage and AUC, respectively. The nonlinearity is small as the correlation coefficients are 0.972 and 0.944 for peak voltage and AUC, respectively. This linear relationship and stable signal amplitude through the experiment demonstrated feasibility of the method as well as little or no loss of beads, DNA, or primers.

This system allowed us to monitor emitted light of known nucleotide incorporation and to determine the reaction stability as well as reproducibility as shown in typical pyrogram in Fig. 2A. The DNA/ primer/SPP complex was subsequently moved into a washing droplet.

The resequencing experiment was repeated five times to suppress random errors and to demonstrate repeatability shown by error bars in Fig. 2B. A mismatched event was performed during the sample loading and unloading to determine the background emission as marked by blue ellipses (Fig. 2C). We found that both the peak voltage as well as the AUC amplitude can be used for identification of the number of incorporated nucleotides. This is confirmed by the histograms in Fig. 2D and E, each containing five groups separated from others. The PMT peak voltage and AUC amplitude are linearly proportional to number of incorporated nucleotides. This linear relationship and stable signal amplitude through the experiment demonstrated feasibility of the method as well as little or no loss of beads, DNA, or primers. Extracted peak voltage and AUC amplitudes for single nucleotide incorporation were (1.667 ± 0.140) V and (0.213 ± 0.026) mC (mean ± standard deviation), respectively (Fig. 2F). The background peak voltage value and AUC amplitudes were only (0.799 ± 0.079) AUC and (0.010 ± 0.012) mC (mean ± standard deviation), respectively, well below the signal for single nucleotide incorporation.

The order of reaction stations for resequencing is designed to match the sequence and incorporate nucleotide(s) in each reaction. Presence of a single nucleotide polymorphism (SNP) is detected as a missing reaction and there will be no nucleotide incorporation once the SNP occurs. The first and last reactions at each glass surface are performed with mismatched nucleotides during sample loading/unloading to detect a background signal and calibrate the system. Once there is a nucleotide mismatch detected during resequencing, the signal will be either stronger than expected (in the case of incorporation of an identical nucleotide), or as intense as the background signal with no incorporation. In the latter case, the pyrosequencing would stop there but the user would know that there is at least one SNP in the tested sequence.

Demonstrating the capability of pyrosequencing method at the glass surface was conducted by performing de novosequencing experiment with a DNA template with the length of 51 bps. The sequence consisted of 16 single, 9 double, 2 triple, 1 quadruple, and 1 quintuple nucleotide incorporations, and 30 mismatches. We performed the experiment with four nucleotides added in a repeated cyclical fashion having 70 reaction stations. This experiment required five glass cover slips and took less than seven hours (Fig. 3A) and was conducted again five times to suppress random errors.

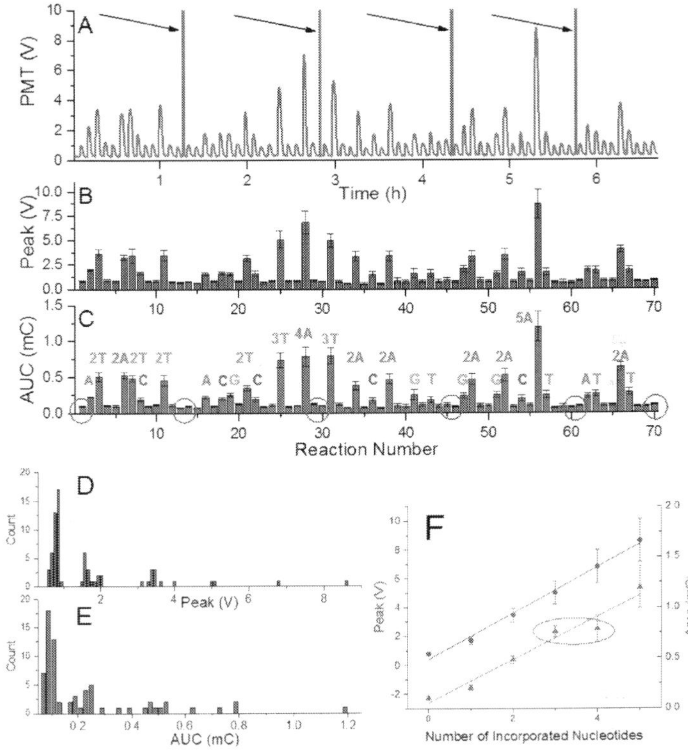

Figure 3: (A) A typical pyrogram of de novo sequencing. (B) Mean value of peak voltage and its standard error as function of reaction number. A mismatched event was performed in the end of the run to determine the background emission (marked by blue ellipses). (C) Mean value of AUC and its standard error as function of reaction. Frequency counts of mean peak values (D) distinguish five groups according to the number of incorporated nucleotides. Mean values of AUC (E) are not as well defined as the ones for peak values. Amplitude of AUC for four nucleotide additions (marked by an ellipse) is higher than the one for five. (F) Peak amplitude (blue circles – left axis) and AUC (red triangles – right axis) as function of number of incorporated nucleotides with slope of (1.605 ± 0.068) V and (0.213 ± 0.022) mC for peak voltage and AUC, respectively. The peak amplitude is linearly proportional to the number of incorporated nucleotides with correlation coefficient of 0.991. The same coefficient for AUC is 0.953. The peak signal amplitude provides more consistent output than the one for AUC.

The captured signals were processed the same way as in the previous experiments (Fig. 3B and C). In addition, histograms of both extracted values (Fig. 3D) show better grouping of the peak voltage amplitude than that for AUCs (Fig. 3E). The signals' peak voltage amplitudes in each group are relatively far from other groups, thus, distinguishing each group is an easy task. The extracted data presented here enabled an accurate sequence determination of all 51 bases. Extracted peak voltage and the AUC amplitude show they are linearly proportional to the number of incorporated nucleotides (Fig. 3F). The value of peak voltage and AUC amplitude are (1.605 ± 0.068) V and (0.213 ± 0.021) mC (mean ± standard deviation), respectively. Again, the peak voltage signal shows better stability than the AUC as marked by blue ellipse at Fig. 3F, where the signal for the four incorporated nucleotides is practically identical to the one for three incorporated nucleotides.

As a final experiment, we tried to incorporate a sequence with length of 81 nucleotides performed in 104 reactions in a single measurement using de novo sequencing configuration. The sequence consisted of 35 single, 8 double, 4 triple, 2 quadruple, and 2 quintuple nucleotide incorporations, and 53 mismatches. The experiment required seven glass cover slips and took less than 10 h (Fig. 4A). Extracted peak voltage and AUC amplitudes were (1.495 ± 0.117) V and (0.182 ± 0.027) mC (mean ± standard deviation), respectively (Fig. 4B and C). The signal per incorporated nucleotide is again well above the background of (0.794 ± 0.117) V and (0.102 ± 0.021) mC (mean ± standard deviation) for peak voltage and AUC amplitudes, respectively (Fig. 4D). The method presented here is then capable of incorporating 81 nucleotides and, as the PMT signal is has consistent amplitude per nucleotide incorporation, there is high probability that even greater number of nucleotide can be incorporated. As observed in previous experiments, peak voltage provides more consistent data than the one of AUC amplitude. The amplitude of AUC for four incorporated nucleotides was higher than the one for 5 nucleotides as marked by arrows in the histograms in Fig. 4E and F.

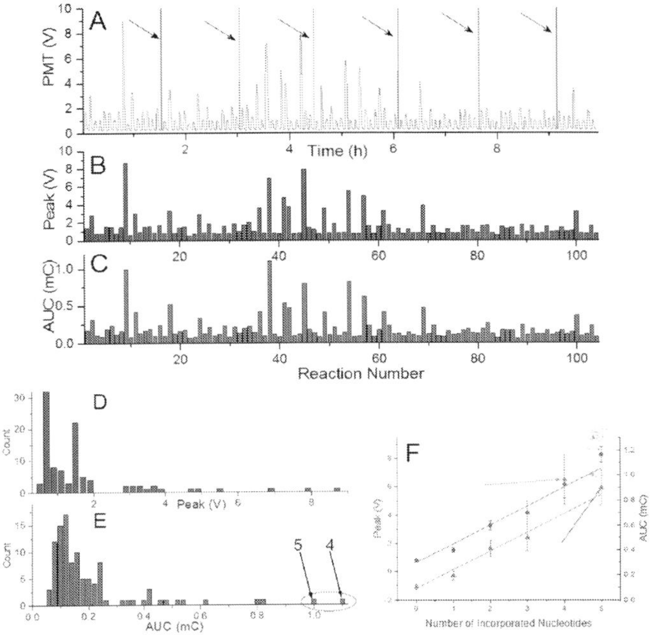

Figure 4: (A) A pyrogram of long de novo sequencing with 104 reactions. (B) Extracted value of peak voltage as function of reaction number. (C) Extracted mean value of AUC as function of reaction number. Frequency counts of mean peak values (D) distinguish six groups according to the number of incorporated nucleotides. (E) Frequency counts of the AUC are not well separated, with signal from four incorporated nucleotides at higher amplitude than the one for five (both marked by arrows). (F) Peak amplitude (blue circles – left axis) and AUC (red triangles – right axis) as function of number of incorporated nucleotides with slope of (1.339 ± 0.145) V and (0.151 ± 0.019) mC for peak voltage and AUC, respectively. The peak amplitude is linearly proportional to the number of incorporated nucleotides with correlation coefficient of 0.944. The same coefficient for AUC is 0.927. The peak signal amplitude then provides more consistent output than the one for AUC.

Overall, we observed that the peak voltage amplitude provided more consistent data than the one of the AUC. The peak voltage amplitude per nucleotide incorporation varied between (1.354 ± 0.075) V, (1.657 ± 0.140) V, (1.605 ± 0.068) V and (1.339 ± 0.145) V (mean ± standard deviation) homopolymeric stretches, re-sequencing (Fig. 2), de novo sequencing of 51 (Fig. 3), and 81 nucleotides (Fig. 4), respectively. The differences are probably caused by slight uncertainty in concentration of chemicals due to manual pipetting as well as temperature fluctuation. Sample preparation was time-consuming and experiments were performed on different days.

CONCLUSION

We have demonstrated a simple pyrosequencing system for both de novo and re-sequencing using open surface microfluidics with magnetic force actuation on a microscope glass cover slip surface without any surface patterning. The droplets were covered with mineral oil, thereby forming virtual reaction chambers. An oil layer was able to sufficiently suppress water evaporation from the sample. As a proof of principle, we designed a robust system with a high signal-to-noise ratio of more than 500 : 1. The ratio of an output signal from PMT to its noise is given by reagents, such as background reactions of substrates, as well as by the ambient light leakage into the PMT system. We observed small, yet constant, light leakage into the system, which affected the PMT output signal.

The peak signal amplitude per nucleotide incorporation was between 1 and 2 V while noise superposed at the signal was below ≈4 mV. The PMT gain was nearly one hundred times lower than its maximum. This large margin in the testing setup suggests that the system has room for improvement, such as increasing its throughput by significantly decreasing the droplet size, probably down to nL level.

We showed a series of experiments starting with re-sequencing providing excellent stable signal with small standard error of the peak voltage amplitude of ≈8.4% per nucleotide incorporation. Loading and unloading stations with on purpose introduced mismatch provided signal with ≈50% amplitude lower than the one with nucleotide incorporation. The system might then become a suitable tool for SNP analysis using this re-sequencing strategy.

We also performed successfully a de novo pyrosequencing experiment with 51 and 81 nucleotide incorporations. The signal degradation with time of number of nucleotide incorporation was minimal; therefore, we can hypothesize that even greater numbers can be incorporated. Increasing the number of sequencing experiments to five suppressed random errors and improved reliability of the method.[37,38]

The system presented here has a flexible configuration that can be used for either de novo sequencing or re-sequencing. These applications can also be altered in real-time based on the instantaneous results. The presented method offers significant advantages in terms of instrument size, simplicity, disposability, and achievable levels of functional integration.

The system should be further optimized to increase the throughput and significantly shorten the time required for each reaction. The process is controlled by the diffusion of individual compounds assisted by simple mixing. Decreasing the sample volume as well as elevating the system's temperature would make the reaction faster and lead to more favorable time required for each pyrosequencing experiment.

An automated pipetting station followed by a lyophilization step also would make this method more user friendly as operators would only require to dispense an appropriate amount of de-ionized water at pre-defined reaction and washing locations and cover those droplets with mineral oil.

The system presented here is compatible with previously designed PCR as well as sample preparation systems. In future, one could create a single complex sample-to-answer system starting with a raw sample, DNA release, its purification, PCR, and pyrosequencing.\

ACKNOWLEDGEMENTS

We would like to thank J. Pipper from Baldr Biotechnologies Pte Ltd, Singapore for useful discussion regarding the pyrosequencing experiment, as well as for his help with the experiment setup and Xun Wenpeng from Northwestern Polytechnic University, Xi'an, China, for his help with MATLAB script. We would like to also to thank Mark Tarn, University of Hull, J, Neužil from Czech Academy of Science and J. Drbohlavová from Brno University of Technology for their editorial help. P. Neužil acknowledges partial financial support by Central European Institute of Technology (CEITEC), grant number CZ.1.05/1.1.00/02.0068.

REFERENCES

1. M. E. W. Saenger, Principles of Nucleic Acid Structure, Springer-Verlag, New York, 1st edn, 1984.
2. L. T. C. França, E. Carrilho and T. B. L. Kist, Q. Rev. Biophys., 2002, 35, 169–200.
3. F. Sanger and A. R. Coulson, J. Mol. Biol., 1975, 94, 441–448.
4. F. Sanger, S. Nicklen and A. R. Coulson, Proc. Natl. Acad. Sci. U. S. A., 1977, 74, 5463–5467.
5. S. D. Boyd, Annu. Rev. Pathol.: Mech. Dis., 2013, 8, 381–410.
6. M. V. Olson, Proc. Natl. Acad. Sci. U. S. A., 1993, 90, 4338–4344.
7. M. L. Metzker, Nat. Rev. Genet., 2010, 11, 31–46.
8. E. M. Harrison, G. K. Paterson, M. T. G. Holden, J. Larsen, M. Stegger, A. R. Larsen, A. Petersen, R. L. Skov, J. M. Christensen, A. Bak Zeuthen, O. Heltberg, S. R. Harris, R. N. Zadoks, J. Parkhill, S. J. Peacock and M. A. Holmes, EMBO Mol. Med., 2013, 5, 509–515.
9. M. Kato-Maeda, C. Ho, B. Passarelli, N. Banaei, J. Grinsdale, L. Flores, J. Anderson, M. Murray, G. Rose, L. M. Kawamura, N. Pourmand, M. A. Tariq, S. Gagneux and P. C. Hopewell, PLoS One, 2013, 8, e58235.
10. C. U. Köser, J. M. Bryant, J. Becq, M. E. Török, M. J. Ellington, M. A. Marti-Renom, A. J. Carmichael, J. Parkhill, G. P. Smith and S. J. Peacock, N. Engl. J. Med., 2013, 369, 290–292.
11. Y. Li and T. O. Tollefsbol, Methods Mol. Biol., 2011, 791, 11–21.
12. A. Alderborn, A. Kristofferson and U. Hammerling, Genome Res., 2000, 10, 1249–1258.
13. P. Nyrén and A. Lundin, Anal. Biochem., 1985, 151, 504–509.
14. M. Ronaghi, S. Karamohamed, B. Pettersson, M. Uhlén and P. Nyrén, Anal. Biochem., 1996, 242, 84–89.
15. E. D. Hyman, Anal. Biochem., 1988, 174, 423–436.
16. M. Ronaghi, M. Uhlen and P. Nyren, Science, 1998, 281, 363–365.
17. A. Ahmadian, M. Ehn and S. Hober, Clin. Chim. Acta, 2006, 363, 83–94.
18. F. Mashayekhi and M. Ronaghi, Anal. Biochem., 2007, 363, 275–287.

19. D. J. Harrison, K. Fluri, K. Seiler, Z. H. Fan, C. S. Effenhauser and A. Manz, Science, 1993, 261, 895–897.
20. M. U. Kopp, A. J. de Mello and A. Manz, Science, 1998, 280, 1046–1048.
21. G. M. Whitesides, Nature, 2006, 442, 368–373.
22. E. K. Sackmann, A. L. Fulton and D. J. Beebe, Nature, 2014, 507, 181–189.
23. D. J. Boles, J. L. Benton, G. J. Siew, M. H. Levy, P. K. Thwar, M. A. Sandahl, J. L. Rouse, L. C. Perkins, A. P. Sudarsan, R. Jalili, V. K. Pamula, V. Srinivasan, R. B. Fair, P. B. Griffin, A. E. Eckhardt and M. G. Pollack, Anal. Chem., 2011, 83, 8439–8447.
24. E. F. Kirkness, Nat. Biotechnol., 2009, 27, 998–999.
25. A. Russom, N. Tooke, H. Andersson and G. Stemme, Anal. Chem., 2005, 77, 7505–7511.
26. B. P. Casavant, E. Berthier, A. B. Theberge, J. Berthier, S. I. Montanez-Sauri, L. L. Bischel, K. Brakke, C. J. Hedman, W. Bushman, N. P. Keller and D. J. Beebe, Proc. Natl. Acad. Sci. U. S. A., 2013, 110, 10111–10116.
27. U. Lehmann, C. Vandevyver, V. K. Parashar and M. A. M. Gijs, Angew. Chem., Int. Ed., 2006, 45, 3062–3067.
28. P. Neuzil, J. Pipper and T. M. Hsieh, Mol. BioSyst., 2006, 2, 292–298.
29. J. Pipper, M. Inoue, L. F. P. Ng, P. Neuzil, Y. Zhang and L. Novak, Nat. Med., 2007, 13, 1259–1263.
30. J. Pipper, Y. Zhang, P. Neuzil and T.-M. Hsieh, Angew. Chem., Int. Ed., 2008, 47, 3900–3904.
31. Z. Long, A. M. Shetty, M. J. Solomon and R. G. Larson, Lab Chip, 2009, 9, 1567–1575.
32. P. Neuzil, W. Sun, T. Karasek and A. Manz, Appl. Phys. Lett., 2015, 106.
33. M. O. Altmeyer, A. Manz and P. Neuzil, Anal. Chem., 2015, 87, 5997–6003.
34. L. Y. Yeo and J. R. Friend, in Annual Review of Fluid Mechanics, Vol 46, ed. S. H. Davis and P. Moin, Annual Reviews, Palo Alto, 2014, vol. 46, pp. 379–406.

35. Z. Guttenberg, H. Muller, H. Habermuller, A. Geisbauer, J. Pipper, J. Felbel, M. Kielpinski, J. Scriba and A. Wixforth, Lab Chip, 2005, 5, 308–317.
36. M. G. Pollack, R. B. Fair and A. D. Shenderov, Appl. Phys. Lett., 2000, 77, 1725–1726.
37. D. Sims, I. Sudbery, N. E. Ilott, A. Heger and C. P. Ponting, Nat. Rev. Genet., 2014, 15, 121–132.
38. M. C. Schatz, A. L. Delcher and S. L. Salzberg, Genome Res., 2010, 20, 1165–1173.

CHAPTER 13

COMPARISON OF NEXT-GENERATION SEQUENCING SYSTEMS

Lin Liu, Yinhu Li, Siliang Li, Ni Hu, Yimin He, Ray Pong, Danni Lin, Lihua Lu, and Maggie Law

NGS Sequencing Department, Beijing Genomics Institute (BGI), 4th Floor, Building 11, Beishan Industrial Zone, Yantian District, Guangdong, China

ABSTRACT

With fast development and wide applications of next-generation sequencing (NGS) technologies, genomic sequence information is within reach to aid the achievement of goals to decode life mysteries, make better crops, detect pathogens, and improve life qualities. NGS systems are typically represented by SOLiD/Ion Torrent PGM from Life Sciences, Genome Analyzer/HiSeq 2000/MiSeq from Illumina, and GS FLX Titanium/GS Junior from Roche. Beijing Genomics Institute (BGI), which possesses the world's biggest sequencing capacity, has multiple NGS systems including 137 HiSeq 2000, 27 SOLiD, one Ion Torrent PGM, one MiSeq, and one 454 sequencer. We have accumulated extensive experience in sample handling, sequencing, and bioinformatics analysis. In this paper, technologies of these systems are reviewed, and first-hand data from extensive experience is summarized and analyzed to discuss the advantages and specifics associated with each sequencing system. At last, applications of NGS are summarized.

INTRODUCTION

(Deoxyribonucleic acid) DNA was demonstrated as the genetic material by Oswald Theodore Avery in 1944. Its double helical strand structure composed of four bases was determined by James D. Watson and Francis Crick in 1953, leading to the central dogma of molecular biology. In most cases, genomic DNA defined the species and individuals, which makes the DNA sequence fundamental to the research on the structures and functions of cells and the decoding of life mysteries [1]. DNA sequencing technologies could help biologists and health care providers in a broad range of applications such as molecular cloning, breeding, finding pathogenic genes, and comparative and evolution studies. DNA sequencing technologies ideally should be fast, accurate, easy-to-operate, and cheap. In the past thirty years, DNA sequencing technologies and applications have undergone tremendous development and act as the engine of the genome era which is characterized by vast amount of genome data and subsequently broad range of research areas and multiple applications. It is necessary to look back on the history of sequencing technology development to review the NGS systems (454, GA/HiSeq, and SOLiD), to compare their advantages and disadvantages, to discuss the various applications, and to evaluate the recently introduced PGM (personal genome machines) and third-generation sequencing technologies and applications. All of these aspects will be described in this paper. Most data and conclusions are from independent users who have extensive first-hand experience in these typical NGS systems in BGI (Beijing Genomics Institute).

Before talking about the NGS systems, we would like to review the history of DNA sequencing briefly. In 1977, Frederick Sanger developed DNA sequencing technology which was based on chain-termination method (also known as Sanger sequencing), and Walter Gilbert developed another sequencing technology based on chemical modification of DNA and subsequent cleavage at specific bases. Because of its high efficiency and low radioactivity, Sanger sequencing was adopted as the primary technology in the "first generation" of laboratory and commercial sequencing applications [2]. At that time, DNA sequencing was laborious and radioactive materials were required. After years of improvement, Applied Biosystems introduced the first automatic sequencing machine (namely AB370) in 1987, adopting capillary electrophoresis which made the sequencing faster and more accurate. AB370 could detect 96 bases one time, 500 K bases a day, and the read length could reach 600 bases. The current model AB3730xl can output 2.88 M bases per day and read length could reach 900 bases since 1995. Emerged in 1998, the automatic sequencing instruments and associated software using the capillary sequencing machines and Sanger sequencing technology became the main tools for the completion of human genome project in 2001 [3]. This project greatly stimulated the development of powerful novel sequencing instrument to increase speed and accuracy, while simultaneously reducing cost and manpower. Not only this, X-prize also accelerated the development of next-generation sequencing (NGS) [4]. The NGS technologies are different from the Sanger method in aspects of massively parallel analysis, high throughput, and reduced cost. Although NGS makes genome sequences handy, the followed data analysis and biological explanations are still the bottle-neck in understanding genomes.

Following the human genome project, 454 was launched by 454 in 2005, and Solexa released Genome Analyzer the next year, followed by (Sequencing by Oligo Ligation Detection) SOLiD provided from Agencourt, which are three most typical massively parallel sequencing systems in the next-generation sequencing (NGS) that shared good performance on throughput, accuracy, and cost compared with Sanger sequencing (shown in Table 1(a)). These founder companies were then purchased by other companies: in 2006 Agencourt was purchased by Applied Biosystems, and in 2007, 454 was purchased by Roche, while Solexa was purchased by Illumina. After years of evolution, these three systems exhibit better performance and their own advantages in terms of read length, accuracy, applications, consumables, man power requirement and informatics infrastructure, and so forth. The comparison of these three systems will be focused and discussed in the later part of this paper (also see Tables 1(a), 1(b), and 1(c)).

Table 1: (a) Advantage and mechanism of sequencers. (b) Components and cost of sequencers. (c) Application of sequencers.

(a)

Sequencer	**454 GS FLX**	**HiSeq 2000**	**SOLiDv4**	**Sanger 3730xl**
Sequencing mechanism	Pyrosequencing	Sequencing by synthesis	Ligation and two-base coding	Dideoxy chain termination
Read length	700 bp	50SE, 50PE, 101PE	50 + 35 bp or 50 + 50 bp	4 0 0 ~ 9 0 0 bp
Accuracy	99.9%*	98%, (100PE)	99.94% *raw data	99.999%
Reads	1 M	3 G	1200~1400 M	—
Output data/ run	0.7 Gb	600 Gb	120 Gb	1.9~84 Kb
Time/run	24 Hours	3~10 Days	7 Days for SE 14 Days for PE	20 Mins~3 Hours
Advantage	Read length, fast	High throughput	Accuracy	High quality, long read length
Disadvantage	Error rate with polybase more than 6, high cost, low throughput	Short read assembly	Short read assembly	High cost low throughput

(b)

Sequencers	**454 GS FLX**	**HiSeq 2000**	**SOLiDv4**	**3730xl**
Instrument price	Instrument $500,000, $7000 per run	Instrument $690,000, $6000/(30x) human genome	Instrument $495,000, $15,000/100 Gb	Instrument $95,000, about $4 per 800 bp reaction
CPU	2* Intel Xeon X5675	2* Intel Xeon X5560	8* processor 2.0 GHz	Pentium IV 3.0 GHz
Memory	48 GB	48 GB	16 GB	1 GB
Hard disk	1.1 TB	3 TB	10 TB	280 GB
Automation in library preparation	Yes	Yes	Yes	No
Other required device	REM e system	cBot system	EZ beads system	No
Cost/million bases	$10	$0.07	$0.13	$2400

(c)

Sequencers	454 GS FLX	HiSeq 2000	SOLiDv4	3730xl
Resequencing		Yes	Yes	
De novo	Yes	Yes		Yes
Cancer	Yes	Yes	Yes	
Array	Yes	Yes	Yes	Yes
High GC sample	Yes	Yes	Yes	
Bacterial	Yes	Yes	Yes	
Large genome	Yes	Yes		
Mutation detection	Yes	Yes	Yes	Yes

(1) All the data is taken from daily average performance runs in BGI. The average daily sequence data output is about 8 Tb in BGI when about 80% sequencers (mainly HiSeq 2000) are running.

(2) The reagent cost of 454 GS FLX Titanium is calculated based on the sequencing of 400 bp; the reagent cost of HiSeq 2000 is calculated based on the sequencing of 200 bp; the reagent cost of SOLiDv4 is calculated based on the sequencing of 85 bp.

(3) HiSeq 2000 is more flexible in sequencing types like 50SE, 50PE, or 101PE.

(4) SOLiD has high accuracy especially when coverage is more than 30x, so it is widely used in detecting variations in resequencing, targeted resequencing, and transcriptome sequencing. Lanes can be independently run to reduce cost.

ROCHE 454 SYSTEM

Roche 454 was the first commercially successful next generation system. This sequencer uses pyrosequencing technology [5]. Instead of using dideoxynucleotides to terminate the chain amplification, pyrosequencing technology relies on the detection of pyrophosphate released during nucleotide incorporation. The library DNAs with 454-specific adaptors are denatured into single strand and captured by amplification beads followed by emulsion PCR [6]. Then on a picotiter plate, one of dNTP (dATP, dGTP, dCTP, dTTP) will complement to the bases of the template strand with the help of ATP sulfurylase, luciferase, luciferin, DNA polymerase, and adenosine 5′ phosphosulfate (APS) and release pyrophosphate (PPi) which equals the amount of incorporated nucleotide. The ATP transformed from PPi drives the luciferin into oxyluciferin and generates visible light [7]. At the same time, the unmatched bases are degraded by apyrase [8]. Then another dNTP is added into the reaction system and the pyrosequencing reaction is repeated.

The read length of Roche 454 was initially 100–150 bp in 2005, 200000+ reads, and could output 20 Mb per run [9, 10]. In 2008 454 GS FLX Titanium system was launched; through upgrading, its read length could reach 700 bp with accuracy 99.9% after filter and output 0.7 G data per run within 24 hours. In late 2009 Roche combined the GS Junior a bench top system into the 454 sequencing system which simplified the library preparation and data processing, and output was also upgraded to 14 G per run [11, 12]. The most outstanding advantage of Roche is its speed: it takes only 10 hours from sequencing start till completion. The read length is also a distinguished character compared with other NGS systems (described in the later part of this paper). But the high cost of reagents remains a challenge for Roche 454. It is about $1 2 . 5 6 × 1 0− 6 per base (counting reagent use only). One of the shortcomings is that it has relatively high error rate in terms of poly-bases longer than 6 bp. But its library construction can be automated, and the emulsion PCR can be semiautomated which could reduce the manpower in a great extent. Other informatics infrastructure and sequencing advantages are listed and compared with HiSeq 2000 and SOLiD systems in Tables 1(a), 1(b), and 1(c).

454 GS FLX TITANIUM SOFTWARE

GS RunProcessor is the main part of the GS FLX Titanium system. The software is in charge of picture background normalization, signal location correction, cross-talk correction, signals conversion, and sequencing data generation. GS RunProcessor would produce a series of files including SFF (standard flowgram format) files each time after run. SFF files contain the basecalled sequences and corresponding quality scores for all individual, high-quality reads (filtered reads). And it could be viewed directly from the screen of GS FLX Titanium system. Using GS De Novo Assembler, GS Reference Mapper and GS Amplicon Variant Analyzer provided by GS FLX Titanium system, SFF files can be applied in multiaspects and converted into fastq format for further data analyzing.

AB SOLID SYSTEM

(Sequencing by Oligo Ligation Detection) SOLiD was purchased by Applied Biosystems in 2006. The sequencer adopts the technology of two-base sequencing based on ligation sequencing. On a SOLiD flowcell, the libraries can be sequenced by 8 base-probe ligation which contains ligation site (the first base), cleavage site (the fifth base), and 4 different fluorescent dyes (linked to the last base) [10]. The fluorescent signal will be recorded during the probes complementary to the template strand and vanished by the cleavage of probes' last 3 bases. And the sequence of the fragment can be deduced after 5 round of sequencing using ladder primer sets.

The read length of SOLiD was initially 35 bp reads and the output was 3 G data per run. Owing to two-base sequencing method, SOLiD could reach a high accuracy of 99.85% after filtering. At the end of 2007, ABI released the first SOLiD system. In late 2010, the SOLiD 5500xl sequencing system was released. From SOLiD to SOLiD 5500xl, five upgrades were released by ABI in just three years. The SOLiD 5500xl realized improved read length, accuracy, and data output of 85 bp, 99.99%, and 30 G per run, respectively. A complete run could be finished within 7 days. The sequencing cost is about \$ 4 0 × 1 0− 9 per base estimated from reagent use only by BGI users. But the short read length and resequencing only in applications is still its major shortcoming [13]. Application of SOLiD includes whole genome resequencing, targeted resequencing, transcriptome research (including gene expression profiling, small RNA analysis, and whole transcriptome analysis), and epigenome (like ChIP-Seq and methylation). Like other NGS systems, SOLiD's computational infrastructure is expensive and not trivial to use; it requires an air-conditioned data center, computing cluster, skilled personnel in computing, distributed memory cluster, fast networks, and batch queue system. Operating system used by most researchers is GNU/LINUX. Each solid sequencer run takes 7 days and generates around 4 TB of raw data. More data will be generated after bioinformatics analysis. This information is listed and compared with other NGS systems in Tables 1(a), 1(b), and 1(c). Automation can be used in library preparations, for example, Tecan system which integrated a Covaris A and Roche 454 REM e system [14].

SOLiD Software

After the sequencing with SOLiD, the original sequence of color coding will be accumulated. According to double-base coding matrix, the original color sequence can be decoded to get the base sequence if we knew the base types for one of any position in the sequence. Because of a kind of color corresponding four base pair, the color coding of the base will directly influence the decoding of its following base. It said that a wrong color coding will cause a chain decoding mistakes. BioScope is SOLiD data analysis package which provides a validated, single framework for resequencing, ChIP-Seq, and whole transcriptome analysis. It depends on reference for the follow-up data analysis. First, the software converts the base sequences of references into color coding sequence. Second, the color-coding sequence of references is compared with the original sequence of color-coding to get the information of mapping with newly developed mapping algorithm MaxMapper.

ILLUMINA GA/HISEQ SYSTEM

In 2006, Solexa released the Genome Analyzer (GA), and in 2007 the company was purchased by Illumina. The sequencer adopts the technology of sequencing by synthesis (SBS). The library with fixed adaptors is denatured to single strands and grafted to the flowcell, followed by bridge amplification to form clusters which contains clonal DNA fragments. Before sequencing, the library splices into single strands with the help of linearization enzyme [10], and then four kinds of nucleotides (ddATP, ddGTP, ddCTP, ddTTP) which contain different cleavable fluorescent dye and a removable blocking group would complement the template one base at a time, and the signal could be captured by a (charge-coupled device) CCD.

At first, solexa GA output was 1 G/run. Through improvements in polymerase, buffer, flowcell, and software, in 2009 the output of GA increased to 20 G/run in August (75PE), 30 G/run in October (100PE), and 50 G/run in December (Truseq V3, 150PE), and the latest GAIIx series can attain 85 G/run. In early 2010, Illumina launched HiSeq 2000, which adopts the same sequencing strategy with GA, and BGI was among the first globally to adopt the HiSeq system. Its output was 200 G per run initially, improved to 600 G per run currently which could be finished in 8 days. In the foreseeable future, it could reach 1 T/run when a personal genome cost could drop below $1 K. The error rate of 100PE could be below 2% in average after filtering (BGI's data). Compared with 454 and SOLiD, HiSeq 2000 is the cheapest in sequencing with $0.02/million bases (reagent counted only by BGI). With multiplexing incorporated in P5/P7 primers and adapters, it could handle thousands of samples simultaneously. HiSeq 2000 needs (HiSeq control software) HCS for program control, (real-time analyzer software) RTA to do on-instrument base-calling, and CASAVA for secondary analysis. There is a 3 TB hard disk in HiSeq 2000. With the aid of Truseq v3 reagents and associated softwares, HiSeq 2000 has improved much on high GC sequencing. MiSeq, a bench top sequencer launched in 2011 which shared most technologies with HiSeq, is especially convenient for amplicon and bacterial sample sequencing. It could sequence 150PE and generate 1.5 G/run in about 10 hrs including sample and library preparation time. Library preparation and their concentration measurement can both be automated with compatible systems like Agilent Bravo, Hamilton Banadu, Tecan, and Apricot Designs.

HiSeq Software

HiSeq control system (HCS) and real-time analyzer (RTA) are adopted by HiSeq 2000. These two softwares could calculate the number and position of clusters based on their first 20 bases, so the first 20 bases of each sequencing would decide each sequencing's output and quality. HiSeq 2000 uses two lasers and four filters to detect four types of nucleotide (A, T, G, and C). The emission spectra of these four kinds of nucleotides have cross-talk, so the images of four nucleotides are not independent and the distribution of bases would affect the quality of sequencing. The standard sequencing output files of the HiSeq 2000 consist of *bcl files, which contain the base calls and quality scores in each cycle. And then it is converted into *_qseq.txt files by BCL Converter. The ELAND program of CASAVA (offline software provided by Illumina) is used to match a large number of reads against a genome.

In conclusion, of the three NGS systems described before, the Illumina HiSeq 2000 features the biggest output and lowest reagent cost, the SOLiD system has the highest accuracy [11], and the Roche 454 system has the longest read length. Details of three sequencing system are list in Tables 1(a), 1(b), and 1(c).

COMPACT PGM SEQUENCERS

Ion Personal Genome Machine (PGM) and MiSeq were launched by Ion Torrent and Illumina. They are both small in size and feature fast turnover rates but limited data throughput. They are targeted to clinical applications and small labs.

Ion PGM from Ion Torrent

Ion PGM was released by Ion Torrent at the end of 2010. PGM uses semiconductor sequencing technology. When a nucleotide is incorporated into the DNA molecules by the polymerase, a proton is released. By detecting the change in pH, PGM recognized whether the nucleotide is added or not. Each time the chip was flooded with one nucleotide after another, if it is not the correct nucleotide, no voltage will be found; if there is 2 nucleotides added, there is double voltage detected [15]. PGM is the first commercial sequencing machine that does not require fluorescence and camera scanning, resulting in higher speed, lower cost, and smaller instrument size. Currently, it enables 200 bp reads in 2 hours and the sample preparation time is less than 6 hours for 8 samples in parallel.

An exemplary application of the Ion Torrent PGM sequencer is the identification of microbial pathogens. In May and June of 2011, an ongoing outbreak of exceptionally virulent Shiga-toxin- (Stx) producing Escherichia coli O104:H4 centered in Germany [16, 17], there were more than 3000 people infected. The whole genome sequencing on Ion Torrent PGM sequencer and HiSeq 2000 helped the scientists to identify the type of E. coli which would directly apply the clue to find the antibiotic resistance. The strain appeared to be a hybrid of two E. coli strains—entero aggregative E. coli and entero hemorrhagic E. coli—which may help explain why it has been particularly pathogenic. From the sequencing result of E. coli TY2482 [18], PGM shows the potential of having a fast, but limited throughput sequencer when there is an outbreak of new disease.

In order to study the sequencing quality, mapping rate, and GC depth distribution of Ion Torrent and compare with HiSeq 2000, a high GC

Rhodobacter sample with high GC content (66%) and 4.2 Mb genome was sequenced in these two different sequencers (Table 2). In another experiment, E. coli K12 DH10B (NC_010473.1) with GC 50.78% was sequenced by Ion Torrent for analysis of quality value, read length, position accuracies, and GC distribution (Figure 1).

Table 2: Comparison in alignment between Ion Torrent and HiSeq 2000

	Ion Torrent[a]	HiSeq 2000[b]
Total reads num	165518	205683
Total bases num	18574086	18511470
Max read length	201	90
Min read length	15	90
Map reads num	157258	157511
Map rate	95%	76.57%
Covered rate	96.50%	93.11%
Total map length	15800258	14176420
Total mismatch base	53475	142425
Total insertion base	109550	1397
Total insertion num	95740	1332
Total deletion base	152495	431
Total deletion num	139264	238
Ave mismatch rate	0.338%	1.004%
Ave insertion rate	0.693%	0.009%
Ave deletion rate	0.965%	0.003%

[a]: use TMAP to align; [b]: use SOAP2 to align

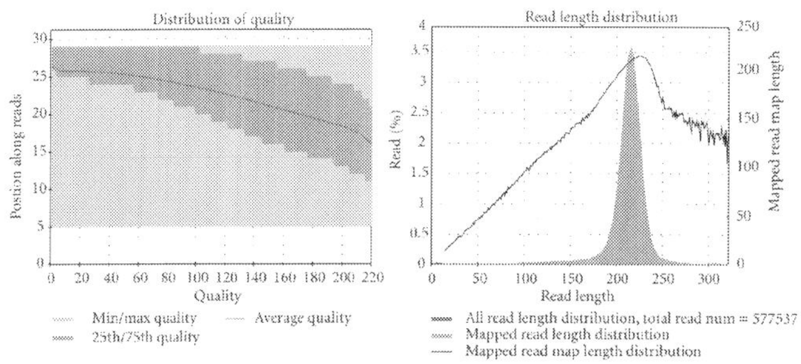

(a)

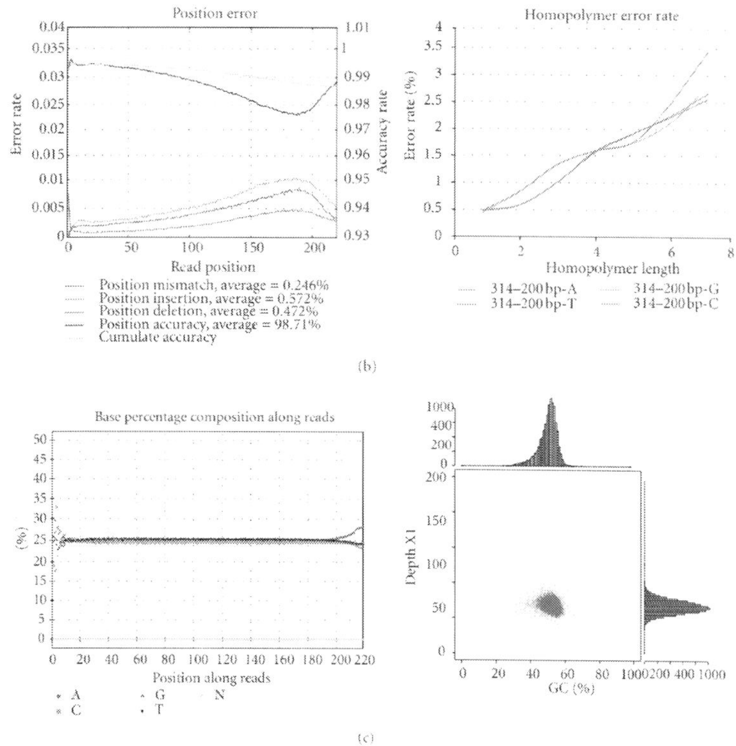

Figure 1: Ion Torrent sequencing quality. E. coli K12 DH10B (NC_010473.1) with GC 50.78% was used for this experiment. (a) is 314–200 bp from Ion Torrent. The left figure is quality value: pink range represents quality minimum and maximum values each position has. Green area represents the top and bottom quarter (1/4) reads of quality. Red line represents the average quality value in the position. The right figure is read length analysis: colored histogram represents the real read length. The black line represents the mapped length, and because it allows 3′ soft clipping, the length is different from the real read length. (b) is accuracy analysis. In each position, accuracy type including mismatch, insertion, and deletion is shown on the left *y*-axis. The average accuracy is shown the right *y*-axis. Accuracy of 200 bp sequencing could reach 99%. (c) is base composition along reads (left) and GC distribution analysis (right). The left figure is base composition in each position of reads. Base line splits after about 95 cycles indicating an inaccurate sequencing. The right one uses 500 bp window and the GC distribution is quite even. The data using high GC samples also indicates a good performance in Ion Torrent (data not shown).

Sequencing Quality

The quality of Ion Torrent is more stable, while the quality of HiSeq 2000 decreases noticeably after 50 cycles, which may be caused by the decay of fluorescent signal with increasing the read length (shown in Figure 1).

Mapping

The insert size of library of Rhodobacter was 350 bp, and 0.5 Gb data was obtained from HiSeq. The sequencing depth was over 100x, and the contig and scaffold N50 were 39530 bp and 194344 bp, respectively. Based on the assembly result, we used 33 Mb which is obtained from ion torrent with 314 chip to analyze the map rate. The alignment comparison is Table 2.

The map rate of Ion Torrent is higher than HiSeq 2000, but it is incomparable because of the different alignment methods used in different sequencers. Besides the significant difference on data including mismatch rate, insertion rate, and deletion rate, HiSeq 2000 and Ion Torrent were still incomparable because of the different sequencing principles. For example, the polynucleotide site could not be indentified easily in Ion Torrent. But it is shown that Ion Torrent has a stable quality along sequencing reads and a good performance on mismatch accuracies, but rather a bias in detection of indels. Different types of accuracy are analyzed and shown in Figure 1.

GC Depth Distribution

The GC depth distribution is better in Ion Torrent from Figure 1. In Ion Torrent, the sequencing depth is similar while the GC content is from 63% to 73%. However in HiSeq 2000, the average sequencing depth is 4x when the GC content is 60%, while it is 3x with 70% GC content.

Ion Torrent has already released Ion 314 and 316 and planned to launch Ion 318 chips in late 2011. The chips are different in the number of wells resulting in higher production within the same sequencing time. The Ion 318 chip enables the production of >1 Gb data in 2 hours. Read length is expected to increase to >400 bp in 2012.

MiSeq from Illumina

MiSeq which still uses SBS technology was launched by Illumina. It integrates the functions of cluster generation, SBS, and data analysis in a single instrument and can go from sample to answer (analyzed data) within a single day (as few as 8 hours). The Nextera, TruSeq, and Illumina's reversible terminator-based sequencing by synthesis chemistry was used in this innovative engineering. The highest integrity data and broader range of application, including amplicon sequencing, clone checking, ChIP-Seq, and small genome sequencing, are the outstanding parts of MiSeq. It is also flexible to perform single 36 bp reads (120 MB output) up to 2 × 150 paired-end reads (1–1.5 GB output) in MiSeq. Due to its significant improvement in read length, the resulting data performs better in contig assembly compared with HiSeq (data not shown). The related sequencing result of MiSeq is shown in Table 3. We also compared PGM with MiSeq in Table 4.

Table 3: MiSeq 150PE data

Sample	GC	Q20	Q30
Human HPV	33.57; 33.62	98.26; 95.52	93.64; 88.52
Bacteria	61.33; 61.43	90.84; 83.86	78.46; 69.04

(1) The data in the table includes both read 1 and read 2 from paired-end sequencing.

(2) GC represents the GC content of libraries.

(3) Q20 value is the average Q20 of all bases in a read, which represents the ratio of bases with probability of containing no more than one error in 100 bases. Q30 value is the average Q30 of all bases in a read, which represents the ratio of bases with probability of containing no more than one error in 1,000 bases.

Table 4: The comparison between PGM and MiSeq

	PGM	**MiSeq**
Output	10 MB–100 MB	120 MB–1.5 GB
Read length	~200 bp	Up to 2 × 150 bp

Sequencing time	2 hours for 1 × 200 bp	3 hours for 1 × 36 single read
		27 hours for 2 × 150 bp pair end read
Sample preparation time	8 samples in parallel, less than 6 hrs	As fast as 2 hrs, with 15 minutes hand on time
Sequencing method	semiconductor technology with a simple sequencing chemistry	Sequencing by synthesis (SBS)
Potential for development	Various parameters (read length, cycle time, accuracy, etc.)	Limited factors, major concentrate in flowcell surface size, insert sizes, and how to pack cluster in tighter
Input amount	µg	Ng (Nextera)
Data analysis	Off instrument	On instrument

Complete Genomics

Complete genomics has its own sequencer based on Polonator G.007, which is ligation-based sequencer. The owner of Polonator G.007, Dover, collaborated with the Church Laboratory of Harvard Medical School, which is the same team as SOLiD system, and introduced this cheap open system. The Polonator could combine a high-performance instrument at very low price and the freely downloadable, open-source software and protocols in this sequencing system. The Polonator G.007 is ligation detection sequencing, which decodes the base by the single-base probe in nonanucleotides (nonamers), not by dual-base coding [19]. The fluorophore-tagged nonamers will be degenerated by selectively ligate onto a series of anchor primers, whose four components are labeled with one of four fluorophores with the help of T4 DNA ligase, which correspond to the base type at the query position. In the ligation progress, T4 DNA ligase is particularly sensitive to mismatches on 3′-side of the gap which is benefit to improve the accuracy of sequencing. After imaging, the Polonator chemically strips the array of annealed primer-fluorescent probe complex; the anchor primer is replaced and the new mixture are fluorescently tagged nonamers is introduced to sequence the adjacent base [20]. There are two updates compared with Polonator G.007, DNA nanoball (DNB) arrays, and combinatorial probe-anchor ligation (cPAL). Compared with DNA cluster or microsphere, DNA nanoball arrays obtain higher density of DNA cluster on the surface of a silicon chip. As the seven 5-base segments are discontinuous, so the system of hybridization-ligation-detection cycle has higher fault-tolerant ability compared with SOLiD. Complete genomics claim to have 99.999% accuracy with 40x depth and could analyze SNP, indel, and CNV with price 5500$–9500$. But Illumina reported a better performance of HiSeq 2000 use only 30x data (Illumina Genome Network). Recently some researchers compared CG's human genome sequencing data with Illumina system [21], and there are notable differences in detecting SNVs, indels, and system-specific detections in variants.

The Third Generation Sequencer

While the increasing usage and new modification in next generation sequencing, the third generation sequencing is coming out with new insight in the sequencing. Third-generation sequencing has two main characteristics. First, PCR is not needed before sequencing, which shortens DNA preparation time for sequencing. Second, the signal is captured in real time, which means that the signal, no matter whether it is fluorescent (Pacbio) or electric current (Nanopore), is monitored during the enzymatic reaction of adding nucleotide in the complementary strand.

Single-molecule real-time (SMRT) is the third-generation sequencing method developed by Pacific Bioscience (Menlo Park, CA, USA), which made use of modified enzyme and direct observation of the enzymatic reaction in real time. SMRT cell consists of millions of zero-mode waveguides (ZMWs), embedded with only one set of enzymes and DNA template that can be detected during the whole process. During the reaction, the enzyme will incorporate the nucleotide into the complementary strand and cleave off the fluorescent dye previously linked with the nucleotide. Then the camera inside the machine will capture signal in a movie format in real-time observation [19]. This will give out not only the fluorescent signal but also the signal difference along time, which may be useful for the prediction of structural variance in the sequence, especially useful in epigenetic studies such as DNA methlyation [22].

Comparing to second generation, PacBio RS (the first sequencer launched by PacBio) has several advantages. First the sample preparation is very fast; it takes 4 to 6 hours instead of days. Also it does not need PCR step in the preparation step, which reduces bias and error caused by PCR. Second, the turnover rate is quite fast; runs are finished within a day. Third, the average read length is 1300 bp, which is longer than that of any second-generation sequencing technology. Although the throughput of the PacBioRS is lower than second-generation sequencer, this technology is quite useful for clinical laboratories, especially for microbiology research. A paper has been published using PacBio RS on the Haitian cholera outbreak [19].

We have run a de novo assembly of DNA fosmid sample from Oyster with PacBio RS in standard sequencing mode (using LPR chemistry and SMRTcells instead of the new version FCR chemistry and SMRTcells). An SMRT belt template with mean insert size of 7500 kb is made and run in one SMRT cell and a 120-minute movie is taken. After Post-QC filter, 22,373,400 bp reads in 6754 reads (average 2,566 bp) were sequenced with the average Read Score of 0.819. The Coverage is 324x with mean read score of 0.861 and high accuracy (~99.95). The result is exhibited in Figure 2.

	Prefilter	Post-QC filter*
Number of bases	84, 110, 272 bp	22, 373, 400 bp
Number of reads	46, 861	6, 754
Mean read length	513 bp	2, 566 bp
Mean read score	0.144	0.819

* MinRL = 50, MinRS = 0.75

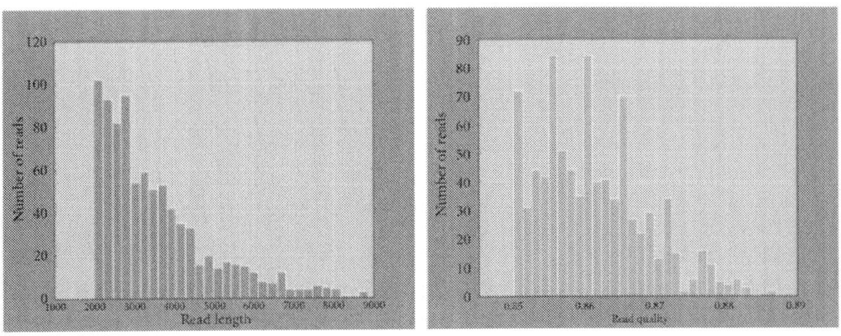

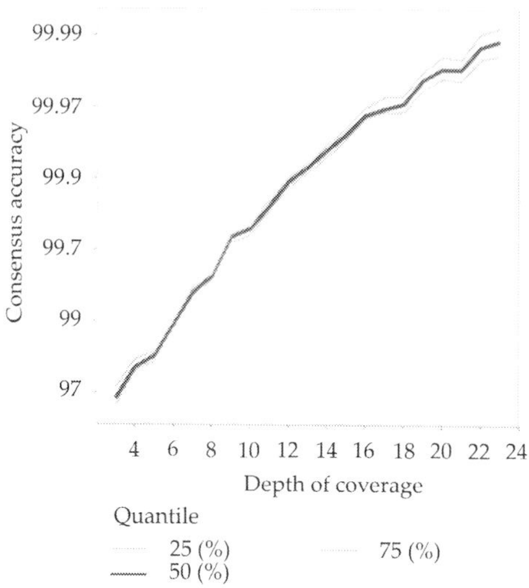

Figure 2: Sequencing of a fosmid DNA using Pacific Biosciences sequencer. With coverage, the accuracy could be above 97%. The figure was constructed by BGI's own data.

Nanopore sequencing is another method of the third generation sequencing. Nanopore is a tiny biopore with diameter in nanoscale [23], which can be found in protein channel embedded on lipid bilayer which facilitates ion exchange. Because of the biological role of nanopore, any particle movement can disrupt the voltage across the channel. The core concept of nanopore sequencing involves putting a thread of single-stranded DNA across α-haemolysin (αHL) pore. αHL, a 33 kD protein isolated from Staphylococcus aureus [20], undergoes self-assembly to form a heptameric transmembrane channel [23]. It can tolerate extraordinary voltage up to 100 mV with current 100 pA [20]. This unique property supports its role as building block of nanopore. In nanopore sequencing, an ionic flow is applied continuously. Current disruption is simply detected by standard electrophysiological technique. Readout is relied on the size difference between all deoxyribonucleoside monophosphate (dNMP). Thus, for given dNMP, characteristic current modulation is shown for discrimination. Ionic current is resumed after trapped nucleotide entirely squeezing out.

Nanopore sequencing possesses a number of fruitful advantages over existing commercialized next-generation sequencing technologies.

Firstly, it potentially reaches long read length >5 kbp with speed 1 bp/ns [19]. Moreover, detection of bases is fluorescent tag-free. Thirdly, except the use of exonuclease for holding up ssDNA and nucleotide cleavage [24], involvement of enzyme is remarkably obviated in nanopore sequencing [22]. This implies that nanopore sequencing is less sensitive to temperature throughout the sequencing reaction and reliable outcome can be maintained. Fourthly, instead of sequencing DNA during polymerization, single DNA strands are sequenced through nanopore by means of DNA strand depolymerization. Hence, hand-on time for sample preparation such as cloning and amplification steps can be shortened significantly.

DISCUSSION OF NGS APPLICATIONS

Fast progress in DNA sequencing technology has made for a substantial reduction in costs and a substantial increase in throughput and accuracy. With more and more organisms being sequenced, a flood of genetic data is inundating the world every day. Progress in genomics has been moving steadily forward due to a revolution in sequencing technology. Additionally, other of types-large scale studies in exomics, metagenomics, epigenomics, and transcriptomics all become reality. Not only do these studies provide the knowledge for basic research, but also they afford immediate application benefits. Scientists across many fields are utilizing these data for the development of better-thriving crops and crop yields and livestock and improved diagnostics, prognostics, and therapies for cancer and other complex diseases.

BGI is on the cutting edge of translating genomics research into molecular breeding and disease association studies with belief that agriculture, medicine, drug development, and clinical treatment will eventually enter a new stage for more detailed understanding of the genetic components of all the organisms. BGI is primarily focused on three projects. (1) The Million Species/Varieties Genomes Project, aims to sequence a million economically and scientifically important plants, animals, and model organisms, including different breeds, varieties, and strains. This project is best represented by our sequencing of the genomes of the Giant panda, potato, macaca, and others, along with multiple resequencing projects. (2) The Million Human Genomes Project focuses on large-scale population and association studies that use whole-genome or whole-exome sequencing strategies. (3) The Million Eco-System Genomes Project has the objective of sequencing the metagenome and cultured microbiome of several different environments, including microenvironments within the human body [25]. Together they are called 3 M project.

In the following part, each of the following aspects of applications including de novo sequencing, mate-pair, whole genome or target-region resequencing, small RNA, transcriptome, RNA seq, epigenomics, and metagenomics, is briefly summarized.

In DNA de novo sequencing, the library with insert size below 800 bp is defined as DNA short fragment library, and it is usually applied in de novo and resequencing research. Skovgaard et al. [26] have applied a combination method of WGS (whole-genome sequencing) and genome copy number analysis to identify the mutations which could suppress the growth deficiency imposed by excessive initiations from the E. coli origin of replication, oriC.

Mate-pair library sequencing is significant beneficial for de novo sequencing, because the method could decrease gap region and extend scaffold length. Reinhardt et al. [27] developed a novel method for de novogenome assembly by analyzing sequencing data from high-throughput short read sequencing technology. They assembled genomes into large scaffolds at a fraction of the traditional cost and without using reference sequence. The assembly of one sample yielded an N50 scaffold size of 531,821 bp with >75% of the predicted genome covered by scaffolds over 100,000 bp.

Whole genome resequencing sequenced the complete DNA sequence of an organism's genome including the whole chromosomal DNA at a single time and alignment with the reference sequence. Mills et al. [28] constructed a map of unbalanced SVs (genomic structural variants) based on whole genome DNA sequencing data from 185 human genomes with SOLiD platform; the map encompassed 22,025 deletions and 6,000 additional SVs, including insertions and tandem duplications [28]. Most SVs (53%) were mapped to nucleotide resolution, which facilitated analyzing their origin and functional impact [28].

The whole genome resequencing is an effective way to study the functional gene, but the high cost and massive data are the main problem for most researchers. Target region sequencing is a solution to solve it.

Microarray capture is a popular way of target region sequencing, which uses hybridization to arrays containing synthetic oligo-nucleotides matching the target DNA sequencing. Gnirke et al. [29] developed a captured method that uses an RNA "baits" to capture target DNA fragments from the "pond" and then uses the Illumina platform to read out the sequence. About 90% of uniquely aligning bases fell on or near bait sequence; up to 50% lay on exons proper [29].

Fehniger et al. used two platforms, Illumina GA and ABI SOLiD, to define the miRNA transcriptomes of resting and cytokine-activated primary murine NK (natural killer) cells [30]. The identified 302 known and 21 novel mature miRNAs were analyzed by unique bioinformatics pipeline from small RNA libraries of NK cell. These miRNAs are overexpressed in broad range and exhibit isomiR complexity, and a subset is differentially expressed following cytokine activation, which were the clue to identify the identification of miRNAs by the Illumina GA and SOLiD instruments [30].

The transcriptome is the set of all RNA molecules, including mRNA, rRNA, tRNA, and other noncoding RNA produced in one or a population of cells. In these years, next-generation sequencing technology is used to study the transcriptome compares with DNA microarray technology in the past. The S. mediterranea transcriptome could be sequenced by an efficient sequencing strategy which designed by Adamidi et al. [31]. The catalog of assembled transcripts and the identified peptides in this study dramatically expand and refine planarian gene annotation, which is demonstrated by validation of several previously unknown transcripts with stem cell-dependent expression patterns.

RNA-seq is a new method in RNA sequencing to study mRNA expression. It is similar to transcriptome sequencing in sample preparation, except the enzyme. In order to estimate the technical variance, Marioni et al. [32] analyzed a kidney RNA samples on both Illumina platform and Affymetrix arrays. The additional analyses such as low-expressed genes, alternative splice variants, and novel transcripts were found on Illumina platform. Bradford et al. [33] compared the data of RNA-seq library on the SOLiD platform and Affymetrix Exon 1.0ST arrays and found a high degree of correspondence between the two platforms in terms of exon-level fold changes and detection. And the greatest detection correspondence was seen when the background error rate is extremely low in RNA-seq. The difference between RNA-seq and transcriptome on SOLiD is not so obvious as Illumina.

There are two kinds of application of epigenetic, Chromatin immunoprecipitation and methylation analysis. Chromatin immunoprecipitation (ChIP) is an immunoprecipitation technique which is used to study the interaction between protein and DNA in a cell, and the histone modifies would be found by the specific location in genome. Based on next-generation sequencing technology, Johnson et al. [34] developed a large-scale chromatin immunoprecipitation assay to identify motif, especially noncanonical NRSF-binding motif. The data displays sharp resolution of binding position (±50 bp), which is important to infer new candidate interaction for the high sensitivity and specificity (ROC (receiver operator characteristic) area ≥0.96) and statistical confidence ($P < 10–4$). Another important application in epigenetic is DNA methylation analysis. DNA methylation exists typically in vertebrates at CpG sites; the methylation caused the conversion of the cytosine to 5-methylcytosine. Chung presented a whole methylome sequencing to study the difference between two kinds of bisulfite conversion methods (in solution versus in gel) by SOLiD platform [35].

The world class genome projects include the 1000 genome project, and the human ENCODE project, the human Microbiome (HMP) project, to name a few. BGI takes an active role in these and many more ongoing projects like 1000 Animal and Plant Genome project, the MetaHIT project, Yanhuang project, LUCAMP (Diabetes-associated Genes and Variations Study), ICGC (international cancer genome project), Ancient human genome, 1000 Mendelian Disorders Project, Genome 10 K Project, and so forth [25]. These internationally collaborated genome projects greatly enhanced genomics study and applications in healthcare and other fields.

To manage multiple projects including large and complex ones with up to tens of thousands of samples, a superior and sophisticated project management system is required handling information processing from the very beginning of sample labeling and storage to library construction, multiplexing, sequencing, and informatics analysis. Research-oriented bioinformatics analysis and followup experiment processed are not included. Although automation techniques' adoption has greatly simplified bioexperiment human interferences, all other procedures carried out by human power have to be managed. BGI has developed BMS system and Cloud service for efficient information exchange and project management. The behavior management mainly follows Japan 5S onsite model. Additionally, BGI has passed ISO9001 and CSPro (authorized by Illumina) QC system and is currently taking (Clinical Laboratory Improvement Amendments) CLIA and (American Society for Histocompatibility and Immunogenetics) AShI tests. Quick, standard, and open reflection system guarantees an efficient troubleshooting pathway and high performance, for example, instrument design failure of Truseq v3 flowcell resulting in bubble appearance (which is defined as "bottom-middle-swatch" phenomenon by Illumina) and random *N* in reads. This potentially hazards sequencing quality, GC composition as well as throughput. It not only effects a small area where the bubble locates resulting in reading *N* but also effects the focus of the place nearby, including the whole swatch, and the adjacent swatch. Filtering parameters have to be determined to ensure quality raw data for bioinformatics processing. Lead by the NGS tech group, joint meetings were called for analyzing and troubleshooting this problem, to discuss strategies to best minimize effect in terms of cost and project time, to construct communication channel, to statistically summarize compensation, in order to provide best project management strategies in this time. Some reagent QC examples are summaried in Liu et al. [36].

BGI is establishing their cloud services. Combined with advanced NGS technologies with multiple choices, a plug-and-run informatics service is handy and affordable. A series of softwares are available including BLAST, SOAP, and SOAP SNP for sequence alignment and pipelines for RNAseq data. Also SNP calling programs such as Hecate and Gaea are about to be released. Big-data studies from the whole spectrum of life and biomedical sciences now can be shared and published on a new journal GigaSicence cofounded by BGI and Biomed Central. It has a novel publication format: each piece of data links to a standard manuscript publication with an extensive database which hosts all associated data, data analysis tools, and cloud-computing resources. The scope covers not just omic type data and the fields of high-throughput biology currently serviced by large public repositories but also the growing range of more difficult-to-access data, such as imaging, neuroscience, ecology, cohort data, systems biology, and other new types of large-scale sharable data.

REFERENCES

1. G. M. Church and W. Gilbert, "Genomic sequencing," Proceedings of the National Academy of Sciences of the United States of America, vol. 81, no. 7, pp. 1991–1995, 1984.
2. http://en.wikipedia.org/wiki/DNA_sequencing/.
3. F. S. Collins, M. Morgan, and A. Patrinos, "The Human Genome Project: lessons from large-scale biology," Science, vol. 300, no. 5617, pp. 286–290, 2003.
4. http://genomics.xprize.org/.
5. http://my454.com/products/technology.asp.
6. J. Berka, Y. J. Chen, J. H. Leamon et al., "Bead emulsion nucleic acid amplification," U.S. Patent Application, 2005.
7. T. Foehlich et al., "High-throughput nucleic acid analysis," U.S. Patent, 2010.
8. http://www.pyrosequencing.com/DynPage.aspx.
9. http://www.roche-applied-science.com/.
10. E. R. Mardis, "The impact of next-generation sequencing technology on genetics," Trends in Genetics, vol. 24, no. 3, pp. 133–141, 2008.
11. S. M. Huse, J. A. Huber, H. G. Morrison, M. L. Sogin, and D. M. Welch, "Accuracy and quality of massively parallel DNA pyrosequencing," Genome Biology, vol. 8, no. 7, article R143, 2007.
12. "The new GS junior sequencer," http://www.gsjunior.com/instrument-workflow.php.
13. "SOLiD system accuray," http://www.appliedbiosystems.com/absite/us/en/home/applications-technologies/solid-next-generation-sequencing.html.
14. http://www.tecan.com/platform/apps/product/index.asp?MenuID=3465&ID=7191&Menu=1&Item=33.52.2.
15. B. A. Flusberg, D. R. Webster, J. H. Lee et al., "Direct detection of DNA methylation during single-molecule, real-time sequencing," Nature Methods, vol. 7, no. 6, pp. 461–465, 2010.
16. A. Mellmann, D. Harmsen, C. A. Cummings et al., "Prospective genomic characterization of the german enterohemorrhagic

Escherichia coli O104:H4 outbreak by rapid next generation sequencing technology," PLoS ONE, vol. 6, no. 7, Article ID e22751, 2011.

17. H. Rohde, J. Qin, Y. Cui et al., "Open-source genomic analysis of Shiga-toxin-producing E. coli O104:H4," New England Journal of Medicine, vol. 365, no. 8, pp. 718–724, 2011.
18. C. S. Chin, J. Sorenson, J. B. Harris et al., "The origin of the Haitian cholera outbreak strain," New England Journal of Medicine, vol. 364, no. 1, pp. 33–42, 2011.
19. W. Timp, U. M. Mirsaidov, D. Wang, J. Comer, A. Aksimentiev, and G. Timp, "Nanopore sequencing: electrical measurements of the code of life," IEEE Transactions on Nanotechnology, vol. 9, no. 3, pp. 281–294, 2010.
20. D. W. Deamer and M. Akeson, "Nanopores and nucleic acids: prospects for ultrarapid sequencing," Trends in Biotechnology, vol. 18, no. 4, pp. 147–151, 2000.
21. "Performance comparison of whole-genome sequencing systems," Nature Biotechnology, vol. 30, pp. 78–82, 2012.
22. D. Branton, D. W. Deamer, A. Marziali et al., "The potential and challenges of nanopore sequencing," Nature Biotechnology, vol. 26, no. 10, pp. 1146–1153, 2008.
23. L. Song, M. R. Hobaugh, C. Shustak, S. Cheley, H. Bayley, and J. E. Gouaux, "Structure of staphylococcal α-hemolysin, a heptameric transmembrane pore," Science, vol. 274, no. 5294, pp. 1859–1866, 1996.
24. J. Clarke, H. C. Wu, L. Jayasinghe, A. Patel, S. Reid, and H. Bayley, "Continuous base identification for single-molecule nanopore DNA sequencing," Nature Nanotechnology, vol. 4, no. 4, pp. 265–270, 2009.
25. Website of BGI, http://www.genomics.org.cn.
26. O. Skovgaard, M. Bak, A. Løbner-Olesen et al., "Genome-wide detection of chromosomal rearrangements, indels, and mutations in circular chromosomes by short read sequencing," Genome Research, vol. 21, no. 8, pp. 1388–1393, 2011.
27. J. A. Reinhardt, D. A. Baltrus, M. T. Nishimura, W. R. Jeck, C. D.

Jones, and J. L. Dangl, "De novo assembly using low-coverage short read sequence data from the rice pathogen Pseudomonas syringae pv. oryzae," Genome Research, vol. 19, no. 2, pp. 294–305, 2009.

28. R. E. Mills, K. Walter, C. Stewart et al., "Mapping copy number variation by population-scale genome sequencing," Nature, vol. 470, no. 7332, pp. 59–65, 2011.
29. A. Gnirke, A. Melnikov, J. Maguire et al., "Solution hybrid selection with ultra-long oligonucleotides for massively parallel targeted sequencing," Nature Biotechnology, vol. 27, no. 2, pp. 182–189, 2009.
30. T. A. Fehniger, T. Wylie, E. Germino et al., "Next-generation sequencing identifies the natural killer cell microRNA transcriptome," Genome Research, vol. 20, no. 11, pp. 1590–1604, 2010.
31. C. Adamidi, Y. Wang, D. Gruen et al., "De novo assembly and validation of planaria transcriptome by massive parallel sequencing and shotgun proteomics," Genome Research, vol. 21, no. 7, pp. 1193–1200, 2011.
32. J. C. Marioni, C. E. Mason, S. M. Mane, M. Stephens, and Y. Gilad, "RNA-seq: an assessment of technical reproducibility and comparison with gene expression arrays," Genome Research, vol. 18, no. 9, pp. 1509–1517, 2008.
33. J. R. Bradford, Y. Hey, T. Yates, Y. Li, S. D. Pepper, and C. J. Miller, "A comparison of massively parallel nucleotide sequencing with oligonucleotide microarrays for global transcription profiling," BMC Genomics, vol. 11, no. 1, article 282, 2010.
34. D. S. Johnson, A. Mortazavi, R. M. Myers, and B. Wold, "Genome-wide mapping of in vivo protein-DNA interactions," Science, vol. 316, no. 5830, pp. 1497–1502, 2007.
35. H. Gu, Z. D. Smith, C. Bock, P. Boyle, A. Gnirke, and A. Meissner, "Preparation of reduced representation bisulfite sequencing libraries for genome-scale DNA methylation profiling," Nature Protocols, vol. 6, no. 4, pp. 468–481, 2011.
36. L. Liu, N. Hu, B. Wang et al., "A brief utilization report on the Illumina HiSeq 2000 sequencer," Mycology, vol. 2, no. 3, pp. 169–191, 2011.

CITATIONS

CHAPTER 6

Català S, Pérez-Sierra A, Abad-Campos P (2015) The Use of Genus-Specific Amplicon Pyrosequencing to Assess *Phytophthora* Species Diversity Using eDNA from Soil and Water in Northern Spain. PLoS ONE 10(3): e0119311. https://doi.org/10.1371/journal.pone.0119311

CHAPTER 7

Kazeeroni EA, Al-Sadi AM. 454-Pyrosequencing Reveals Variable Fungal Diversity Across Farming Systems. *Frontiers in Plant Science.* 2016;7:314. doi:10.3389/fpls.2016.00314.

CHAPTER 8

G. Fusco a, M.G. Amoroso, N. Gesualdi Montesano, and M. Viscardi, Development of a pyrosequencing assay for the typing of alphaherpesviruses, Published online 2015 Jan 12. doi: 10.1016/j.mex.2015.01.001.

CHAPTER 9

Fabrice A, Didier R (2009) Exploring Microbial Diversity Using 16S rRNA High-Throughput Methods. J Comput Sci Syst Biol 2:074-092. doi:10.4172/jcsb.1000019

CHAPTER 10

Shavanthi Rajatileka, Karen Luyt, Maggie Williams, David Harding, David Odd, Elek Molnár, and Anikó Váradi1, Detection of three closely located single nucleotide polymorphisms in the EAAT2 promoter: comparison of single-strand conformational polymorphism (SSCP), pyrosequencing and Sanger sequencing, Published online 2014 Jul 5. doi: 10.1186/1471-2156-15-80.

CHAPTER 11

Zongli Zheng, Abdolreza Advani, O¨ jar Melefors, Steve Glavas, Henrik Nordstro¨ m, Weimin Ye1 , Lars Engstrand and Anders F. Andersson, Titration-free massively parallel pyrosequencing using trace amounts of starting material, doi:10.1093/nar/gkq332.

CHAPTER 12

Ana V. Almeida, Andreas Manz and Pavel Neužil, Pyrosequencing on a glass surface, DOI: 10.1039/c6lc00114a.

CHAPTER 13

Lin Liu, Yinhu Li, Siliang Li, et al., "Comparison of Next-Generation Sequencing Systems," Journal of Biomedicine and Biotechnology, vol. 2012, Article ID 251364, 11 pages, 2012. doi:10.1155/2012/251364

INDEX

S

T

V

Z